G R E E N L A N D

I C E L A N D

Arctic Circle

Ellesmere I.

Thank God Harbour

Smith Sound

Jones Sound

Wellington Channel

Ellis I.

Devon I.

Barrow St.

Lancaster Sound

Somerset I.

Fury Beach

Prince Regent Inlet

Boothia Pen.

G. of Boothia

Melville Pen.

Prince Charles I.

Foxe Basin

Baffin Island

Baffin Bay

Disko I.

Davis St.

C. Farewell

Cumberland Sound

Southampton I.

C. Fullerton

Hudson St.

Frobisher Bay

Resolution I.

C. Chidley

A T L A N T I C   O C E A N

Hudson Bay

Ungava Bay

800 km.

500 mi.

JBG85

ARCTIC
WHALERS
ICY SEAS

ARCTIC WHALERS ICY SEAS

Narratives of the Davis Strait Whale Fishery

IRWIN PUBLISHING
Toronto Canada

W. GILLIES ROSS

Canadian Cataloguing in Publication Data
Ross, W. Gillies (William Gillies), 1931-
    Arctic whalers, icy seas

Includes index.
ISBN 0-7725-1524-7

1. Whaling – Davis Strait – History. 2. Davis
Strait – Description and travel. I. Title.
SH382.2.R67 1985   917.163'42   C85-099048-3

Illustration Credits
Frontispiece: *Men of the American 311-ton whaling bark* Black Eagle *laying out lines
to ice anchors and hauling a whaleboat over the ice. The painting is by Charles E. Allen,
master of the* Black Eagle *on wintering voyages to Cumberland Sound and Hudson Bay
in 1860-61 and 1862-63. (Peary-MacMillan Arctic Museum and the Arctic Studies
Center, Bowdoin College)*
On copyright page (top): *Harpoon gun. (Hull Museums)*
On epigraph page: *Wooden headboard of a sailor's grave at Niantelik, near the former
Cumberland Sound whaling station at Blacklead Island. (Gil Ross)*
page 60: *Original sketches from the journal of John Wanless. (Dundee Museums)*
The following illustration credits have been shortened in the text
and are here given with the full credit line:
Dundee Museums: Dundee Museums and Galleries, Scotland
Hull Central Library: Humberside County Library Service
Hull Museums: Town Docks Museum, Hull Museums and Art Galleries
Glasgow University Library: The Librarian, Glasgow University Library
Metro Toronto Library: Metropolitan Toronto Library Board
The MIT Museum: Hart Nautical Collection, The MIT Museum
National Library of Scotland: The Trustees of the National Library of Scotland

Design by The Dragon's Eye Press
Typeset by Fleet Typographers Ltd.
Maps by Jonathan Gladstone/J.B. Geographics
Printed in Canada by John Deyell Co.
1 2 3 4 5 6 7 8 JD 92 91 90 89 88 87 86 85
Published by Irwin Publishing Inc.

But let them remember that ice is stone,
A floating rock in the stream,
A promontory or an island when aground,
Not less solid than if it were
A land of granite.

JOHN ROSS, *Narrative of a second*
voyage in search of a North-West
Passage (London: A.W. Webster, 1835)

Contents

Acknowledgements

This volume has grown out of studies on the nature and distribution of bowhead whaling activities in the Canadian Arctic and the influence of those activities on biological resources and native population. Grants in support of the work have been obtained at one time or another from the Canada Council, the National Museum of Man (Ottawa), the Department of Indian Affairs and Northern Development (Ottawa), the Canadian Research Centre for Anthropology (Ottawa), the National Advisory Committee on Geographical Research (Ottawa), and Bishop's University (Lennoxville, Quebec). Although the grants were not intended specifically for the preparation of this book they contributed in a significant way to its conception, and the assistance of those institutions is therefore gratefully acknowledged. I extend sincere thanks to the following individuals, whose assistance and cooperation has been indispensable: Shirley Cass (Vienna); Arthur Credland (Keeper, Hull Museums and Art Galleries); Mr. and Mrs. William Cruden (Aberdeen); Kate Christie Gordon-Rogers (former editor, *Polar Record*); David Henderson (Assistant Keeper, Natural History, Dundee Museums and Art Galleries); Dr. Dorothy Johnston (Assistant Archivist, Aberdeen University Library); J.F. Kidd (Archivist, Social and Cultural Archives, Public Archives of Canada); Michael Livingstone-Learmonth (Michinhampton, Gloucester); Vernon Nelson (Archivist, Moravian Archives, Pennsylvania); and J.S. Richie (former Keeper, National Library of Scotland). The staff of the Scott Polar Research Institute, Cambridge, have been helpful and hospitable far beyond the call of duty during more than one visit. Special thanks are due the Humberside Leisure Services Department and all the institutions in which the unpublished narratives used in this book exist. I wish to thank Pauline Callon Clarke, Ursula Thomas, and Jane Whiting for typing parts of the manuscript, and Mieke Köppen Tucker for translating related material from the Dutch. Lastly, I acknowledge the generosity of Bishop's University, whose grant of a study leave in 1980−81 enabled the project to be brought to completion. Finally, I wish to express my gratitude to Rick Archbold for his constructive suggestions and careful editorial scrutiny.

Acknowledgement of the sources of the illustrations reproduced in the book can be found in the "Note on Sources" on page 254.

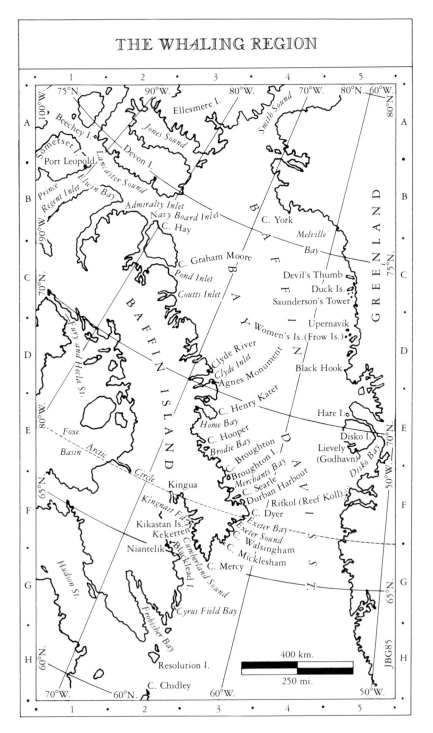

THE WHALING REGION

Introduction

The commercial hunt for the Greenland or bowhead whale *(Balaena mysticetus)* began in earnest off Spitsbergen early in the seventeenth century and spread relentlessly westward to the Greenland Sea, Davis Strait, Baffin Bay, and Hudson Bay. During the nineteenth century the hunt was taken up in arctic waters off the northwestern extremity of North America, in the Bering, Chukchi, and Beaufort seas. The traditional arctic whaling industry continued until the early twentieth century, when the scarcity of whales and a declining demand for oil and baleen brought it to an end.

Behind this extraordinary circumpolar quest was the need of European and American homes, cities, and factories, for illuminants and lubricants before the ascendancy of petroleum products and electricity. By the middle of the nineteenth century whaling had grown to massive proportions. It involved thousands of ships, tens of thousands of men, and innumerable jobs and facilities in related industrial and commercial activities in the principal ports. Although the arctic whale fishery was only part of this vast economic enterprise spanning the world's oceans it was very important because it provided baleen (popularly called "whalebone" or "bone"), springy, filter-feeding slabs with hair-like fringes that exist only in the mouths of the bowhead and other baleen whales. Whalebone was, of course, the principal structural element in the highly engineered clothing of fashionable ladies, but in addition to its well-known use in corset stays and skirt hoops, it had dozens of less glamorous applications in such commonplace articles as umbrella ribs, fishing rods, riding crops, whips, brushes, sieves, nets, window blinds and gratings, chair bottoms and backs, upholstery stuffing, carriage springs, luggage, and fences.

The largest and most enduring whale fishery off arctic North America was that of the Davis Strait – Baffin Bay region, which persisted for more than two centuries prior to the First World War. It was based on what appears to have been a discrete stock of bowhead whales, whose annual cycle of feeding, reproduction, and migration was largely confined to the waters between Greenland and arctic Canada. From a wintering zone northeast of Labrador the main body of whales moved northward along the west Greenland coast in April, May, and June, then crossed Baffin Bay westward wherever they could find a passage through the so-called Middle Ice to enter Lancaster Sound and adjacent waters during June. Other whales were able to penetrate the

Middle Ice in the south to reach Cumberland Sound and other localities along the coast of Baffin Island. The furthest limits of the summer feeding range occurred in Smith Sound, Jones Sound, Barrow Strait, and Prince Regent Inlet. In late August the whales began the return migration, moving southward along the east coast of Baffin Island, and to a lesser extent along the Greenland coast, to their winter habitat among the broken pack ice south of Davis Strait, bordering the open water of the north Atlantic. The entire region through which the whales migrated, and the whaling operations that were carried out there, were generally known as the Davis Strait whale fishery, although they extended far beyond the specific geographic feature of that name.

In the late seventeenth century and through the eighteenth the Davis Strait stock of whales suffered from continuous hunting pressure by Danish, Dutch, German, American, and British whalemen. Were it not for the fact that this whaling was confined to the "East Side"—along the Greenland coast—the whales might have been exterminated in this early period. In retrospect it seems that a part of the stock must have been able to avoid the whaling fleet in summer, possibly by migrating closer to the coast of Baffin Island than the whaleships were inclined to venture, and that this part of the stock acted as a genetic reservoir to perpetuate the population; for by 1825, after more than a century of intense exploitation, there were still about 11,000 whales in the region, according to a recent estimate. But by 1820 the cetacean riches of the "West Side" had become known to the whaling masters, and the entire emphasis of the whaling quickly shifted to Lancaster Sound, the coast of Baffin Island, and the waters of Cumberland Sound.

During the subsequent century—the period covered by this book—whaleships made a total of over 2,300 voyages to the Davis Strait whaling region. More than 92% of these were made by British vessels, approximately 7% by American ships, and the rest by Dutch, French, and German ships.

To the whaling interests of Great Britain the Davis Strait whale fishery of the nineteenth century represented a significant geographical expansion of a long-established system of renewable-resource use. It resulted in a short period of higher yields after the waters of the West Side were first exploited in 1820, but the larger catches were made at the cost of longer voyages to and from the whaling grounds, and higher losses in ships and men owing to the severity of ice conditions in the region. The initial boom was followed by declining yields as the whale population was steadily reduced by uncontrolled killing.

For the American whaling industry Davis Strait whaling provided an alternative to sperm whaling in tropical and temperate waters of the world. The Davis Strait region was relatively close to New England ports, so that voyages could be shorter in duration than those for sperm whales, which sometimes lasted three or four years. Even so, comparatively few American vessels participated in the arctic fishery.

For the colony that would become the Dominion of Canada in 1867 arctic whaling meant little. Sovereignty over the Arctic Islands was uncertain; no Canadian ships were involved; the whale products and profits went elsewhere; indeed most of the people in the settled southern regions knew nothing at all about the enterprise. Nonetheless, along with the Atlantic cod fishery, the fur trade, the forest industry, and mining, whaling was one of the great economic forces to affect the vast area that is now Canada. The expansion of the whaling frontier to the coastal zones of the eastern Arctic in 1820 had important biological, cultural, and political consequences. The pursuit of bowheads into the farthest reaches of their geographical range led to the reduction of the stock to the few hundred animals believed to exist today. The wintering of British and American ships in arctic harbours profoundly affected Eskimo life. The unrestrained activities of whalemen contributed to the transfer of the Arctic Islands from Great Britain to Canada and the extension of Canadian authority into the region.

The arctic whaling industry concentrated so much on the bowhead whale that whalemen, in their journals, simply referred to it as a "whale," or even a "fish." When they had occasion to mention other species they employed more precise terms. Narwhal and beluga, which had been taken by native hunters for centuries, became popular targets for the harpoons of commercial whalemen in the second half of the nineteenth

A swivel harpoon gun mounted on the bow of a whaleboat. Harpoons such as this one were used during the nineteenth century, especially by the British, because they had a greater range than the standard hand-thrown harpoon. (Dundee Museums)

century as the number of bowheads declined from overhunting. Bottlenose whales were exploited as well. After the bowhead, the beluga or white whale was probably the most important species for European whalers. It is a comparatively small animal, roughly twenty feet in length, and it can be secured in large numbers by herding and netting. The bowhead by contrast is a mighty animal running to fifty-five feet or more in length, thirty feet around the waistline, sixty tonnes or more in weight, with tail flukes measuring more than twenty feet from tip to tip, and a mouth large enough to accommodate a dozen Jonahs. Far from being the "ignoble Leviathan" pursued by "inferior souls," as Herman Melville would have us believe, the bowhead is a majestic creature and the whalemen who hunted it were inferior to none. These brave men sailed in wooden hulls among floes and bergs, stalked the great whales on arctic seas whose temperatures never rose much above freezing, pulled the oars of open whaleboats and plunged in harpoons and lances with fingers benumbed by cold, all the while pushing the navigation season beyond its reasonable limits and sometimes getting trapped within the congealing ice pack of winter. Writers who experienced the hazards of the arctic whale fishery knew the truth of the matter.

According to William Scoresby, Jr., whose whaling achievements and published writings made him the best-known of all Greenland whaling masters, the men "were exposed to danger from three sources, viz. from the ice, from the climate, and from the whales themselves." Thus the arctic whalemen faced two challenges more than their counterparts in the tropics; and arctic ice and climate were certainly not adversaries to trifle with.

But let the reader judge. In this volume nineteenth century whalemen relate their own experiences at the Davis Strait whale fishery. These fifteen firsthand accounts by whaling masters and surgeons, a mate, a boatsteerer, and a sportsman-adventurer are presented chronologically. In the text my introduction, conclusion, and occasional interpolated commentaries to each chapter appear in a type size slightly smaller than the type size of the original selections.

Whaling is today a controversial and emotion-charged topic in which the disparate views of industrialists, biologists, conservationists, and others are often heard. In offering to the public a collection of nineteenth-century whaling narratives it is not my intention to stimulate a debate on the morality of killing whales commercially. I do, however, remark later in the book upon the failure of the traditional whaling industry to perceive that unlimited pressure upon the whales could lead only to destruction of a renewable resource and the demise of the industry itself. Regardless of what we may feel about the ethics of killing animals or about the short-sighted policy of unrestricted whaling during the nineteenth century we can not fail to admire and be moved by the skill, fortitude, endurance, and courage of the men who actively sought the bowhead whale in the northern seas.

A NOTE ON THE EDITING

The chapters that have been published before (see "Note on Sources") have undergone cosmetic changes only, in order to bring their spelling and format into line with other selections. But the manuscript journals, compiled on board ship under difficult conditions, usually written in pencil, often in small handwriting or hurried scrawl with home-made abbreviations to conserve scarce paper or time, and frequently faded a century or so later, presented many problems and required more editing. The authors, I felt, would resent any extensive modification of their writings, but they would probably wish them to be tidied up into a form acceptable for publication and comprehensible to readers. Therefore I have not tampered with word choice, other than to correct some uncomfortably bad past participles such as "I seed a whale," and I have not altered syntax, but I have adjusted capitalization, punctuation, and spelling to conform with modern usage and create a reasonable degree of consistency. I have standardized the form of journal entries, and have expanded many abbreviations, with the exception of terms normally used to summarize ship's position and weather at the end of a daily entry, such as "lat." (latitude), "long." (longitude), "NW" (northwest), and so on. Place names remain in the forms used by British and American whalemen. For Greenland I have not attempted to give Danish or modern Greenlandic equivalents. The locations should be clear from the accompanying maps.

A word in square brackets is one that was missing in the original manuscript but which I have supplied to complete an author's statement or make a meaning clear. If such a word is followed by a question mark it denotes a word that was in the original manuscript but cannot be identified with certainty. A totally unreadable word is indicated as "illegible." My deletion of accidentally repeated words or other material is represented by three dots. Explanations of unfamiliar terms (including those relating to sea ice, whaling, and other nautical matters) can be found in the glossary at the end of the book, although a number of them are explained in brackets following their first occurrence in the text.

In the original selections I have left the words "Eskimo" and "Inuit" as used by each of the various writers. In my own text, however, I have used throughout the accepted English term "Eskimo," whose usage extends over three centuries. Aside from being widely understood it can be used as adjective as well as noun, and as it avoids regional differences in native self-designation (*Inuit, Inuvialuit, Kalâlek,* and so on) it can usefully denote the native people of both Greenland and arctic Canada, who are referred to in this book. Whatever the origin of the word "Eskimo" it has no derogatory connotations in my mind, nor did it, I believe, in those of nineteenth century whalemen.

ARCTIC WHALERS
ICY SEAS

Davis Strait whaling was carried out in an age of sail. Steam propulsion was not introduced until 1857, and then only as an auxiliary to sail power. The steam whaler Maud, *with all sails set, attempts to make way through dense pack ice, 1889. Photo by Walter Livingstone-Learmonth. (Public Archives Canada, C-88303)*

One   This Immense Body of Ice

...the ship was plyed to windward between the land and ice which now approached within eight or ten miles and even in that space was straggling ice and icebergs. Within an area of six or eight miles I accounted 108 icebergs... W.E. CASS, 1824

ALTHOUGH WE MARVEL TODAY at the navigational capabilities of large ice-breakers and super-tankers, whose power and momentum can drive them through sea ice with apparent ease, we know that even such impressive modern machines as these have their limitations. The gigantic *Manhattan*, a tanker as long as three football fields, with engines developing 43,000 shaft horsepower, was stopped several times and was holed by ice in arctic waters less than two decades ago. If some people are now concerned about the vulnerability of modern ore-carriers and tankers, what would they think of the nineteenth-century whaleships—mere cockleshells by comparison—that sailed to the Davis Strait whale fishery? Their hulls were made of wood, not steel. Many were under 150 feet in length, approximately as long as a supertanker is broad, and had capacities of a few hundred tons—utterly minuscule when compared to the *Manhattan*'s 115,000 deadweight tons. Furthermore, most of the whalers used in the nineteenth century, having no engines, had to rely almost entirely upon the vagaries of the wind.

These vessels sailed year after year to the arctic seas without any of the conveniences of radio, radar, sonar, air observation, or ice and weather forecasting. It is hardly a surprise that they paid a price for their boldness. One out of every seventeen voyages, on the average, ended in the loss of the vessel. Ice was only one of several formidable natural hazards confronting the arctic whalemen, but it was unquestionably the most dangerous.

On a voyage to the Davis Strait whaling grounds the whalemen usually encounterd three types of floating ice: icebergs, pack ice, and fast ice. Of the three types, icebergs were the most spectacular. Immense, deep-draught monoliths of land ice calved into the sea from the sinuous arms of the massive Greenland ice sheet and other glaciers situated around the Baffin Bay region, icebergs drift along with the slow but purposeful movement of oceanic currents, first northward along the west coast of Greenland, then westward across Baffin Bay, and finally southward down the coast of Baffin Island towards Newfoundland and the destructive warmth of the Gulf Stream. European whalemen would usually first sight these majestic ice castles off Cape Farewell, the southern tip of Greenland, in April or May, but there would be plenty of opportunities later on to inspect them at closer quarters, for the normal summer itinerary of the whalers resembled the drift pattern of the icebergs.

Pack ice consists of innumerable floes of sea ice of various sizes and shapes, drifting in response to complex forces generated by currents and winds. In winter virtually all of Davis Strait and Baffin Bay is covered with ice of this type, which tends to be carried southward by the Labrador Current to form a fringe all along the coast to southeastern Newfoundland, a thousand miles from Hudson Strait. It was at the outer limit of this impressive land-hugging barrier of ice between latitudes 60°N and 63°N, in the region known as the "Southwest Ice," that the ships carried out their whale fishing early in the season. Later, as they managed to penetrate more northerly waters with the advance of the "warm" season, they were close neighbours with the Davis Strait and Baffin Bay pack through the months of May and June, and not infrequently July and August as well.

The pack ice was sometimes open enough to sail through, and was then known as "loose ice," "sailing ice," or "open pack"; but at other times it could be jammed tightly together to form an impenetrable barrier. There was no question of intentionally cutting a lane through solid ice by the force of the ship because sea ice after a winter's growth is about six feet thick, and much thicker where pressure ridges exist. Nor was it feasible – especially with wooden sailing ships – to penetrate the pack by simply pushing the floes aside. Ice floes, even relatively small ones, are dangerous adversaries, "floating rocks in the stream," as Sir John Ross called them, which can easily hole a ship's hull. Generally low in relief, the arctic pack ice is much less spectacular in appearance than the tall, forbidding icebergs, but it is in fact more dangerous, owing to its great extent and its awesome power when driven by wind.

After a few weeks of "fishing" (whaling) at the Southwest Ice in April most of the ships would proceed northward along the pack ice edge, following the whales towards Davis Strait. This course converged with the northwest-trending coast of Greenland, and as the ships drew in with the land they encountered a third type of ice. Landfast ice, fast ice, or as the whalemen often referred to it, the "land floe," consists of sea ice attached firmly to the shore. In winter and spring fast ice fills the fiords and bays, and projects some distance beyond the head-lands to its outer limit, the "floe edge." The fast ice is relatively stable and immobile, in contrast to the pack ice, which is always in motion.

When the whaleships arrived in April or May, pack ice and fast ice were at their maximum geographical extent. During subsequent months the seaward margin of the landfast ice would gradually disintegrate and retreat towards the coasts, while ablation of the countless floes of the drifting pack would steadily reduce their size and create more open water among them. By mid-August in most years the whaling grounds were relatively clear of ice, and captains could cruise where they wished.

In many ways the ice could be helpful and comforting. The floe edge provided a natural deep-water wharf where vessels could moor easily and the surface of the landfast ice, which tended to be relatively flat and undisturbed, constituted a convenient platform for the movement of people and goods between ship and shore. Whaling vessels often hooked on to the floe edge while their whaleboats cruised, or while dead whales were being processed; and when the pressures of encroaching pack ice threatened the ships, they could be warped into man-made "docks" hurriedly cut into the fast ice. Drinking water could be obtained from small ponds on icebergs or from old floes, in which downward migration of brine crystals gradually turned salt water ice into fresh. During gales, whaling masters sometimes sought refuge in the calmer waters within the pack ice, and they frequently moored to the more protected lee sides of large floes and bergs.

Yet, under the enormous pressures generated by persistent winds upon a body of pack ice, the same useful and protective ice floes could become rock-hard battering-rams capable of crushing a 300-ton whaler to matchwood in minutes. The individual floes, once given horizontal and vertical motion by wind, waves, and currents, could be formidable hazards, especially when hidden by darkness, fog, rain, or snow. Icebergs, although they covered far less area than pack ice, were nonetheless a very severe danger, their mass giving them an irresistible momentum when propelled by currents. In the open sea, if sighted in time, they could be avoided, but when they drifted through tightly

Icebergs were numerous in Davis Strait and Baffin Bay. Although dangerous adversaries in times of poor visibility, they were also useful as sources of fresh water. (Glasgow University Library)

packed floes among which ships were beset and powerless to move, any contact between iceberg and wooden hull spelled disaster.

The distribution of ice strongly influenced the movements of the whalers. In a typical whaling voyage a ship would arrive at the Southwest Ice in the spring, sail northward along the eastern margin of the pack ice to Disko Bay, follow any shore lead that opened up between the Baffin Bay pack ice and the fast ice of the Greenland coast, penetrate the dreaded pack of Melville Bay to reach the open North Water, which extended south from Smith Sound, and arrive at the mouth of Lancaster Sound in early July. After whaling in this region for a month or two the vessel would normally cruise southeastward along the coast of Baffin Island and take her departure homeward from Davis Strait.

The whaler *Brunswick* departed from Hull for the Davis Strait whaling on 17 March 1824, in company with five other ships. Among them was the *Jane* (see chapter 5) and also the *Isabella*, which six years earlier had carried John Ross on the voyage of discovery that initiated the great revival of the British quest for the Northwest Passage. Nine years later, in 1833, by the sheerest of coincidences, the *Isabella* would be in Lancaster Sound at precisely the right time and place to rescue the men of John Ross's second expedition (generally assumed dead since their departure from England more than four years before) as they straggled out of the Arctic Islands in a few open boats, "unshaven since I know not when, dirty, dressed in the rags of wild beasts instead of the tatters of civilization, and starved to the very bones," their hopes for survival hanging upon the slim chance of meeting a whaling vessel that had remained longer than usual in the region.

A model of the *Brunswick* shows that she was a three-masted ship (square-rigged on all three masts), broad of beam, and like most whalers built for strength and cargo capacity rather than speed. Details reveal her unequivocally as a whaler: a blubber guy extends from foremast to mainmast well above the deck, and from it hang two speck tackles, or blocks for hoisting blubber from whales made fast alongside; and seven whaleboats hang from davits, three boats on each side of the ship and one at the stern. Contemporary records give the *Brunswick*'s tonnage as 357 and her year of construction as 1814. There is little doubt that like other vessels of the arctic whaling fleet her hull was fortified internally, and sheathed externally with hardwood or iron at the bows and along the waterline, to withstand the pressure and abrasive action of ice – "strengthened and doubled" as sailors called it.

In the previous season Captain Blyth had taken this squat, unglamourous vessel some 1500 miles across the north Atlantic before the winter's end, navigated her another 2500 miles or so through the icy regions of Baffin Bay, and sailed her back to Hull as a "full ship" containing the oil and bone of 36 whales – an extraordinary catch.

English and Scottish whalers normally departed from mainland ports with their crews only half made-up and completed their rosters with men from the Orkney or Shetland islands. These tough men, tempered against the rigours of northern climates, and well-skilled in the handling of small boats, made ideal whalemen. Accordingly, the *Brunswick* dropped anchor in the harbour at Stromness, Orkney, on 21 March, and Captain Blyth signed on twenty-three additional crew. With a total of fifty he would be able to dispatch seven five-oared whaleboats in pursuit of whales, with enough men left over to handle the ship.

On 23 March the *Brunswick* weighed anchor and headed west for Davis Strait.

A week out of Stromness supplies were sold to the men, including "stockings, mittens, flannel shirts, and jackets, apparel as are commonly worn in the polar regions," as well as "bibles and testaments." With both physical and spiritual comfort thus assured, attention was next turned to the preparation of whaling gear and the initiation of green hands into the fraternity of arctic whalemen. In a traditional ceremony, Neptune's assistants lathered the faces of the initiates with a mixture of tar and grease and shaved them with a monstrous, jagged-edge "razor" of hoop iron. On 8 April the lofty east coast of Greenland was sighted and the ship soon rounded Cape Farewell and headed north into Davis Strait, forgoing the fishing at the Southwest Ice.

William Eden Cass, twenty-two-year-old son of a Yorkshire country doctor, and himself a medical student, had shipped as surgeon on the *Brunswick*. It was not his first arctic whaling voyage, for he had evidently sailed to Davis Strait in 1821 and may have returned again in 1822, but his earlier experiences do not appear to have dulled his enthusiasm or his intense curiosity about polar phenomena. It is clear that he had read contemporary accounts of the northern regions and, like most whaling surgeons, he kept a shipboard journal of observations, events, and impressions during the voyage. Cass later rewrote part of his original journal, and it is from his revised account that the selection is taken.

After the whaling voyage he describes below, Cass completed his medical training at Guys Hospital, London, and became a member of the Royal College of Surgeons in 1826. In 1837 he married a young lady of Goole, close to his own birthplace of Howden, and spent the rest of his life practising medicine there. In 1887 he was honoured by the Good Samaritan Lodge of Oddfellows for fifty years of service as surgeon to the lodge, and for his association "with every movement having for its object the alleviation of the suffering and distressed, and notably and heroically so, during severe visitations of cholera." Cass died in 1889 at at the age of 88.

Following the usual itinerary of whalers entering Davis Strait, the *Brunswick* sailed north, converging on the coast of Greenland, called by the sailors the East Land (as distinct from the West Land on the opposite side of Davis Strait and Baffin Bay). Cass found the scenery "sublime." Behind barren snow-covered islands and the rugged mainland coast rising steeply from the water were "stupendous, terrific and craggy mountains." But what most excited his admiration and interest was the ice in the

Far left: Sail-propelled whalers, such as the Hull ship Brunswick *shown here, sailed to the Davis Strait whale fishery from a dozen or more ports in England and Scotland in the early nineteenth century. (Hull Museums)*

Left: Steam auxiliary whalers were more typical of late nineteenth century arctic whaling, which was dominated by the Scottish ports of Dundee and Peterhead. (Dundee Museums)

sea. He wrote about it with a comprehensiveness normally lacking in the journals and logbooks compiled by professional whalemen, whose familiarity with ice led to stoic acceptance of it. Old hands seldom bothered to set down geographical or scientific facts about ice; to them it was simply an accepted part of any arctic whaling voyage. To William Eden Cass, however, the ice was fascinating and entirely deserving of comment.

During the winter the surface of the water in Davis Strait becomes frozen over except in such parts as from the continued agitation of the water, the freezing process does not gain an ascendancy. It is found by experience that this immense body of ice approximates with the East Land no further south than the sixty-seventh degree of latitude. From thence [southward] it gradually recedes from the East Land to the West Land and the Straits become wider, forming a space of water between the ice and the East Land down to Cape Farewell of considerable magnitude.

As spring advances this immense body of ice extending from the head of Baffin Bay down to the Labrador coast and to a very low latitude and from the East to the West Land is sent asunder by the united efforts of wind and tide. It is impossible that an undulating sea can reach far beyond the margin of such a compact body of ice into fields and floes. (By a field of ice is to be understood a piece of ice the limit of which cannot be seen from a ship's masthead.) As summer advances these fields and floes are in a state of incipient decay; and as an undulating sea reaches them, they are soon reduced apparently to a small fraction of their original bulks, so that in particular seasons by the end of September the greater part of the body of ice which occupies Davis Strait in the early part of spring is annihilated....

In the early part of the season the ice contiguous to the East Land from the latitude of 67°N to 70°N breaks up, and what is not destroyed by the constant friction of the small pieces together, drifts to the westward...so that generally by the end of May the ice lies upwards of fifty miles from the land in the latitudes of 68°N and 69°N. By the latter part of June ships are generally able to sail with the greatest facility to the latitude of 75°N, where they generally meet with considerable obstruction.

It is probable that the ice which approximates with the West Land does not break up so early as the ice near the East Land. After it is broken up on the West side and at the head of Baffin's Bay, the ice lays in the centre of Davis Strait and Baffin Bay and becomes a movable body which, during the prevalence of different winds, alters greatly its position....

These fields and floes of ice vary in thickness and in circumference. They are often seen possessing an area of twelve or fourteen square miles; the floes that are generally seen are from two to six or seven miles in circumference and vary from three feet to forty feet in thickness. Some of them have a level surface for miles whilst others are extremely rugged, probably from being produced by the cohesion of small pieces, the remains of the floes which have been broken up.

It is probable that the most terrific and sublime spectacle in nature is the concussion of these enormous fields and floes. It would indeed be difficult for the human imagination to conceive anything more awful and impressive than the sensations produced on the minds of the crew of one solitary ship working her way through the regions of eternal frost under a dark and lurid atmosphere and the sun obscured by dense vapours, when the still and utter silence which had reigned around is suddenly and fearfully interrupted by the meeting of two enormous fields or floes revolving in opposite directions and advancing against each other at the rate of several miles an hour. The one is broken and destroyed or forced in part above the other with a loud and terrible dissonance, resembling the voice of thunder or the roaring of cannon. During this terrible contact huge masses of ice are raised with tremendous force above the surface of the water and projected upon

the further surface of the superincumbent field. These disrupted masses are known under the name of hummocks and are very common in Davis Strait, being thrown up to the height of twenty feet from the surface of the field extending fifty or sixty yards in length and forming a mass of about 2,000 tons in weight.

The amazing changes which take place in the position of the most compact ice are truly singular; they even astonish those accustomed to their occurrence. Thus ships immovably fixed with regard to the ice have been known to perform a complete revolution in a few hours; and two ships beset a few furlongs apart within the most compact ice have sometimes been separated to the distance of several leagues within the space of two or three days, notwithstanding that

the apparent continuity of the ice remained unbroken. The motion of ice is occasioned chiefly by currents or the pressure of other ice. The wind also has the effect of driving all ice to leeward with a velocity nearly in the inverse proportion to its depth under water. Light ice consequently drives faster than heavy ice, and loose ice than fields.

The effect of polar ice on the climate and the phenomena of the atmosphere are considerable. It affects the colour of the sky, diminishes the violence of the wind and equalizes the temperature of the air. Thus a storm will frequently blow on one side of a field for a considerable time before it becomes perceptible on the other. As a ship approaches the ice in a gale of wind, it greatly diminishes and a few miles within a body of ice, it is often little wind; in short I have seen ships a few miles within a body of ice with all their sails set whilst the ships without were under double-reefed topsails.

As I have said before, wind has a great effect upon ice, in driving it to leeward; southerly winds have the property of driving ice with amazing velocity to the northwards. I have seen a floe upward of ten miles in circumference and drawing thirty to forty—and in some parts fifty—feet of water, driven during a southerly wind in the course of five days, thirty miles. Northerly winds drive the ice to the south, and of course, tend to clear the country of ice.... Immediately above these congregated ices, is a beautiful and natural effulgence of light, visible at a great distance and even in the darkest night during the prevalence of the densest fog, named the ice blink.

Pack ice was the nemesis of arctic whaleships. When floes pressed together and rafted one upon the other under the influence of wind, the pack formed an impassable barrier and threatened a ship's survival. (Dundee Museums)

MODERN ICE OBSERVERS and glaciologists classify sea ice according to various characteristics visible from aircraft or by remote sensing. With the much more limited vantage point of a small ship (and the dinner-table conversation of Captain Blyth and his officers) surgeon Cass was able to distinguish between giant floes and small ones, to recognize hummocks and other variations in surface morphology, to discern the influence of both winds and currents in imparting movement to the floes, to note the frictional drag of pack ice texture on wind velocity, and to observe the phenomenon of ice blink (ice reflection from low clouds). Of course, none of this was new to the whalemen. Their survival on every voyage depended largely on their ability to "read" the weather and the ice, and to take prompt evasive or defensive action when these powerful antagonists conspired against them. Their senses were continually tuned to subtle variations in wind strength and direction, cloud cover, visibility, and the concentration of pack ice. Even when off watch and in their bunks there was no relaxation. They slept fitfully, constantly alert for the slightest change in the rhythm of small ice fragments tapping their way along the hull, jarred violently awake from time to time by the shock of the ship's stem meeting a large piece head-on.

Most of the thousands of icebergs produced each year in the Davis Strait–Baffin Bay region are calved (broken off) from the snouts of glaciers which advance into the sea along the Greenland coast. This fact was known at the time but the theory mentioned below by Cass, that the glaciers themselves were formed within valleys adjacent to the coast by local accumulation of snowfall, was incorrect. To shipboard observers, whose vision did not extend beyond the coastal mountains and glaciers, the hypothesis was not unreasonable, but later in the nineteenth century, when European explorers and alpinists ventured inland, they discovered that the coastal glaciers were in fact nourished by a vast ice sheet. This massive body of ice, the largest in the northern hemisphere, covers roughly three-quarters of a million square miles and reaches a thickness of almost two miles. Snow falling on its higher elevations is gradually transformed by the pressure of younger layers accumulating above into glacier ice, whose basal portions move imperceptibly towards the coasts. The journey from initial snowfall to iceberg production takes an estimated three thousand years, so the coastal glaciers are not, as Cass suggests, "as old as the earth itself."

The most sublime and splendid spectacle in northern scenery is the magnificent and stupendous mountains of ice, denominated icebergs, which are interspersed with the West Ice, and are seen afloat and aground in great numbers in Davis Strait and Baffin Bay. They are as various in size as in figure; their summits are from twenty to 300 feet and their circumferences from three miles to 200 or 300 yards. Enormous as these dimensions appear, the part which projects above the surface of the water amounts only to a sixth part of the whole mass, the proportion of the mass above water being generally estimated at one foot to six below. Some of these icebergs are seen representing islands rising by a gradual declivity to their summit and then terminating abruptly in a perpendicular, craggy, and impending cliff. The surfaces of these are broken by mountains of no mean size with deep valleys between. Sailors during the time the ships were beset in 1821 often amused themselves, when they were able to gain the top of such bergs, by sliding down the mountains on a piece of board into the valleys below.

The eye of fancy may see a thousand fantastic figures in these icebergs, exhibiting the ruined forms of castles and cathedrals with their walls, gates, towers, and spires in every state of decay, with the different objects of animated nature. Some of these masses are so completely perforated that a boat may sail through them. When the sun shines full upon them, the prospect is inconceivably brilliant, assuming all the various hues and tints that the reflection of the solar orb on their rude surfaces can convey. Their lustre is too dazzling of the eyes, and the air is filled with astonishing brightness. These icebergs, it is supposed, are

separated from mountains of ice which occupy the narrow inlets of Baffin Bay, and which are of an immense height and extend considerably beyond the precipices to which they are attached[,] into the sea. Captain Ross distinguishes them by the name of glaciers; one he... mentions as filling a space of four miles square and extending one mile into the sea, its height being at least 1000 feet. These glaciers probably derive their origin from an accumulation of snow which is swept by the violence of the storm from the adjoining hills into the valleys and inlets below, filling them to the tops of the cliffs, when by frequent thaws and the occasional fall of moisture interrupting the frost during the first part of the winter, the snow will in some small degree dissolve, by which means it only acquires a greater hardness when the frost returns, and during the course of that rigorous season becomes a very compact body of ice. The nucleus of some of these glaciers may be as ancient as the earth itself, being replenished and increased by the congelation of snow.

These beautiful icy cliffs are, in common with every species of ice, very fragile during the summer months. They frequently, by the weight of superincumbent snows, &c., assume an overhanging form, and are precipitated into the seas. Water also, by its expansion in secret cavities

during the period of freezing, frequently detaches these icebergs with tremendous force. They are thus and by other means converted into floating bergs or ice islands.... The ice of which bergs are composed is remarkably hard, compact, and semi-transparent, affording excellent water. During the summer water runs in tremendous currents down the sides of the bergs from whence the whaling ships are generally supplied, the ice being liquefied by the sun.

WHALING MASTERS HAD TO KNOW about the characteristics and distibution of ice in northern waters; the survival of their ships and crews often depended on their knowledge and judgement. There were, however, no training courses and few books on sea ice and glaciers. For whaling captains, experience was the great teacher, combined with information passed orally from one veteran to another both on the whaling grounds and at home ports. The whaleman's practical knowledge of arctic ice was far ahead of the formal "scientific" knowledge in Europe and America, and in fact the first comprehensive description of sea ice was written by a whaling captain, William Scoresby, Jr., who had begun his career in the Greenland whale fishery (between Greenland and Spitsbergen) at the age of ten and risen to his own command by twenty-two. His two-volume *Account of the Arctic Regions*, published in 1820, outlined the characteristics of the arctic regions and the history and methods of the arctic whaling industry. Almost a fifth of the first volume was devoted to a discussion of ice—its types, properties, origin, distribution, movement, and impact on whaling activities. Scoresby's pioneer treatise became widely read and frequently quoted. Its influence is clearly discernible in Surgeon Cass's narrative above, which contains at least three sentences lifted directly from that source. Much of the surgeon's description, however, is based on his own experiences on the Davis Strait whaling grounds, a region to which William Scoresby, Jr., never had the privilege to sail.

The men of the Dundee *work desperately with ice saws to save their ship as the* Harlingen *(background) is crushed by the pressure of the pack ice. The artist's rendition of icebergs within the pack is, to say the least, fanciful. The sketch comes from the published version of Duncan's narrative.*

Two

We Seemed to Stand Alone

We seemed to stand alone in the frozen ocean, the devoted victims of its rage and fury.
DAVID DUNCAN, 1827

MANY SHIPS WERE LOST in the course of arctic whaling, and most of them were crushed by ice. The awe-inspiring force of ice floes and icebergs could threaten the safety of vessels in any part of the Davis Strait whaling grounds, but Melville Bay, on the northwest coast of Greenland, earned a reputation for ship destruction that was unsurpassed. Ordinarily the whalers found an early passage northward along the coast of Greenland in a progressively narrowing lead between the land floe on their right hand and the so-called "Middle Ice" of Baffin Bay on their left, but when winds shifted and blew from the west, the pack ice was forced in against the floe edge. For ships sailing along it, there was no refuge other than the crude "docks" the men frantically cut with their twelve- or fourteen-foot ice saws. Inevitably some whaleships were nipped and crushed against the firm bulwark of the landfast ice by the tremendous pressure of the advancing pack.

It was in this perilous situation that the English whaler *Dundee* and the Dutch whaler *Harlingen* found themselves on August 1826. As the following account shows, only the *Dundee* escaped destruction. Her companion vessel, only a few hundred yards away, was pushed over and crushed in short order by the grinding ice.

The tale is told by the captain of the *Dundee*, David Duncan, who had by then spent twenty years as master of arctic whalers. Duncan was a careful observer, whose remarks on newly discovered features along the east Greenland coast, and the names he had applied to them, had been published a few years before in a scientific journal. His straightforward narrative needs no dramatic embellishment. The facts, simply stated, speak for themselves in this early account of a winter drift in the arctic pack ice by a British whaleship and her crew.

The *Dundee* was a 358-ton ship registered in London, strengthened and doubled for polar voyages, of which she was already a veteran. On 3 April she departed from London and sailed north to Lerwick in the Shetland Islands where, in the usual custom of whaling vessels, she recruited additional men, bringing her complement up to forty-eight. On the eighteenth she was under way again, and a month later the crew sighted the first icebergs and floes about 200 miles east of the mouth of Hudson Strait. This was the Southwest fishery—a productive whaling ground until the whales were fished out, but also a "dangerous fishery," as Scoresby had emphasized, exposed to the strong gales, high seas, and low temperatures of a still-wintry North Atlantic, and hindered at the outset by long hours of darkness that forced the boats to carry lanterns. Whales were seen on 19 May and the first kill was made on the twenty-fourth, a very large bowhead with baleen of 11 feet, 6 inches. The next month was spent plying north along the edge of the pack, occasionally venturing into "straggling ice," and pursuing whales at any hour of the day or night—a strenuous and demanding routine that may have in some way contributed to the unexplained death of two of the Shetlandmen.

In late June the *Dundee* reached Disko Bay, about half way up the Greenland coast, and received from the whaler *William* the first intimations of a bad ice year. The *Cicero* of Hull had already been crushed and the *William* had some of the survivors on board. Two of them, John Thompson and Peter Slater, were transferred to the *Dundee* to ease the strain upon the *William's* food stores. One can only guess what impression their vivid stories of shipwreck must have made upon the seamen of the *Dundee*, freshly arrived on the whaling grounds. And one can only wonder at the irony that these two men, who were fortunate indeed in being rescued from a sinking ship, were destined to be carried by the *Dundee* into far more dreadful circumstances within the arctic pack.

North of Disko, the *Dundee* made slow progress, sailing wherever possible and, when ice or calm interfered, advancing yard by yard in tow behind her own whaleboats or pulled by her sailors tracking in harness along the land floe like beasts of burden. On 7 July, at latitude 72°N, forty-four ships were in sight, all attempting to force a way northward along the floe edge toward Melville Bay. The work was arduous but the scene often colourful, gay, and even sonorous, as the marching and pulling chants of hundreds of men wafted over the still water and ice.

But time was running out. In other seasons whalers had often managed to reach the mouth of Lancaster Sound before mid-July, but as this July drew to a close the ships were still hemmed in by ice on the Greenland coast. Among the captains, hopes of making a successful passage through Melville Bay had dwindled and some had already turned their vessels southward to probe at lower latitudes for a route through the Middle Ice. Captain Duncan was one of a small number of masters who continued to persevere towards the north. His decision was unfortunate, for it was to result in a winter imprisonment amid the floes of Baffin Bay and Davis Strait, but the shipmasters of those days had pitifully little information on which to base their decisions. Experience, intuition, visual observation of weather and ice from deck and crow's nest, and the news from other ships were all they had to go on.

The following selection is from a small book written by Captain David Duncan after the voyage. It reveals him as a rational and effective commander who faced extraordinary hazards without panic and described disastrous events without excessive emotion. Was he really the cool, detached observer and self-confident leader that he appears to be? No one can know. Whatever he confided to his original shipboard journal is now lost. Our impressions of his personality and leadership, and of the occurrences of the voyage, must necessarily be based upon his subsequent published account.

The most constant companion of the *Dundee* in late July and through most of August was the Dutch whaler *Harlingen*, commanded by Captain Klass Hoekstra of the island of Texel. A century earlier, in 1726, more than a hundred Dutch ships had sailed to the Davis Strait whale fishery but now there were only three, and these the last representatives of Dutch enterprise in the region.

---

1 AUGUST 1826

Made fast to an iceberg; could not proceed any further northward for ice; saw one of the Duck Islands, bearing east.

2 AUGUST

Ice closing very fast; let go from the berg, and drifted away with the loose ice; fell in with the *Harlingen*, of

Harlingen, Captain Hockstra [Hoekstra]; unshipped the rudder to prevent its being injured by ice.

3 AUGUST

Ice beginning to slacken; shipped the rudder; set all sail possible, and bored the ship through the ice into a small hole of water, to the NNW; latter part plying to windward and at times laying to. *Harlingen* in company.

4 AUGUST

Dodging in a hole of water among heavy ice....

6 AUGUST

Latitude 73°28′N. Run to the south west among a great quantity of loose ice, there being no appearance of a passage through to the north west at present. *Harlingen* still in company....

8 AUGUST

Latitude 72°54′N, longitude 54°W. This day we fell in with the *Ariel*, of Hull, Captain Watson, who informed us that he had been as far to the southward as 71°00′N, and no appearance of any passage through to the westward, that he had seen several vessels and they were all running further to the southward; so we agreed to accompany the *Ariel* and *Harlingen*, and return back again in hopes of procuring a northerly passage towards the West Land, as the wind had prevailed a little time from the south east....

10 AUGUST

Boring to the north east among a great quantity of loose ice. *Dundee* and *Harlingen* made fast to a small floe of ice, finding it impossible to get any further. *Ariel* beset about a mile from us.

12 AUGUST

Ice coming down the northward at a tremendous rate; unshipped the rudder, and slacked round to the lee side of the floe. *Ariel* still lying beset; *Harlingen* made fast beside us. At this time there were a great many icebergs around us.

13 AUGUST

Shipped the rudder, and reached the ship into a small hole of water; ice closing very fast; made the ship fast to an iceberg. *Harlingen* in company; the *Ariel* lying beset with ice, about six miles from us....

16 AUGUST

Latitude 73°N, longitude 63°W. All hands employed sawing the ice all round the ship, in order to form a dock for her. *Harlingen* in company. The immense labour and fatigue attending this operation may be easily imagined, when it is considered that the ice was full seven feet in thickness, and nearly as hard as granite. The crew of the *Harlingen* also endeavoured to saw a dock in like manner, but the floe was so thick and heavy, they could not succeed....

19 AUGUST

Close beset among heavy ice. Three sails in sight. No water seen from masthead....

22 AUGUST

Called all hands and began to saw a dock into the ice. Ship lying beset among floes, the ice running very fast; our ship in great danger.

23 AUGUST

Lying close beset; all hands employed sawing the ice. And here I have to relate the afflicting loss of our companion the *Harlingen*. About two A.M. I observed the *Harlingen*'s ensign hoisted in distress, and sent a party of our men over the ice to their assistance; but before they got a ship's length, the *Harlingen* fell on her broadside, occasioned by two large floes of ice meeting together. The people,

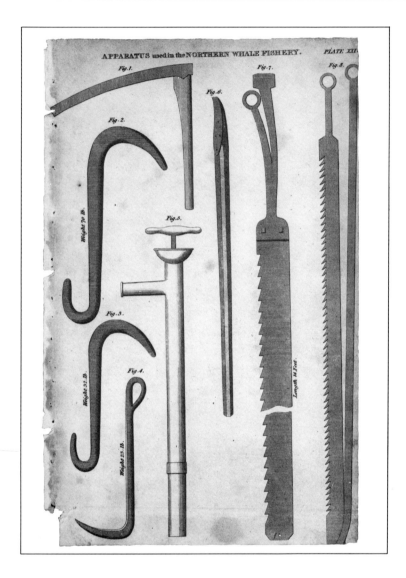

Ice apparatus illustrated in An Account of the
Arctic Regions, *by William Scoresby, Jr. Fig. 1 —*
ice axe; fig. 2, 3 — ice anchors; fig. 4 — bay ice anchor;
fig. 5 — blubber pump; fig. 6 — bone hand spike;
fig. 7 — ice saw; fig. 8 — a saw, with moveable back,
for thinner ice ("the back part serves to keep the
lower extremity forward, and to preserve the saw in
a perpendicular position"). (Metro Toronto Library)

forty-six in number, came on board the *Dundee*, and were
all readily and kindly received by myself and our whole
crew, and treated in every respect as our own people.
Captain Hockstra, his son, and nephew, chief mate, surgeon,
and a boy shared the cabin with me. The *Harlingen* was an
entire wreck, and the accident so very sudden, that the
men could save but very little even of their clothes.

24 AUGUST
Our ship lying, for the present, safe in the dock we had
made in the ice; our whole attention was given to the
wreck, to get every thing that was possible from her; but
part of the vessel being under water, the carpenters were
employed cutting away the deck, in hopes of getting at
some of the provisions, but it proved useless. Saw one
whale this day, but it disappeared immediately afterwards.

25 AUGUST
Watch employed sawing the ice; and both ship's [ships']
companies endeavouring to save what they could from the
wreck, which was still lying under water....

27 AUGUST
Latitude 74°01′N. Finished our dock, after sawing five
hundred feet into the ice. Put the Dutchmen on the same

Far left: This sketch, one of many by the famous Hull whaling master William Barron, depicts whalemen "putting a blast in the ice."
(Hull Museums)

Left: The Devil's Thumb, a well known landmark that marked the entrance to dreaded Melville Bay. One of a number of pencil, pen, and watercolour sketches illustrating a journal by R. H. Hilliard, surgeon of the Dundee whaler Narwhal *in 1859.*
(Glenbow Archives)

allowance as the English. Saw the Devil's Thumb, bearing ESE, distance about thirty-six miles. The making of a dock is thus performed. The floe of ice is on its edge, sawed inwards in a triangular form, of a greater length and width than the ship, and then by various cross cuts, the whole of the ice is forced out and forms the dock; and thereby sufficient room is made to receive the ship and let her float therein, preventing the pressure of the floes on the ship, and all immediate danger from the approach of ice bergs. This herculean task is done with saws from fourteen to sixteen feet long....

29 AUGUST

Latitude 73°57′N, longitude 62°30′W. Bay ice making very fast; employed at the wreck, got some sails and a few broken spars. No water to be seen from the masthead.

30 AUGUST

People employed trying to heave the wreck upright, in hopes of saving some of the provisions, but all their efforts were in vain. Breaking the bay ice round our own ship.

31 AUGUST

The frost was now very severe; watch employed breaking the ice round the ship; put all hands on short allowance;

very little water seen from the masthead; some of the hands employed at the wreck endeavouring to get at some of the provisions.

1 SEPTEMBER

Employed at the wreck, and fired one rocket down the main hatchway, in hopes of blowing the decks up; employed breaking the bay ice round the ship; procured from the wrecked ship one cask of pork, a forestay and two warps....

6 SEPTEMBER

Close beset, tracking the wreck towards the ship....

8 SEPTEMBER

Latitude 73°56′N, longitude 63°W. Drying the sails belonging to the wreck; about seven P.M. got the wreck round to the *Dundee*; rigged purchases in expectation of heaving her up, but the ice kept too firm a hold of her, and baffled all our endeavours; very little water seen from the masthead. Thermometer 14°F below zero.

10 SEPTEMBER

Employed at the wreck; no water seen from the masthead; saw the Devil's Thumb bearing SSE distance about sixty miles....

13 SEPTEMBER

At three A.M. called all hands; began to saw further into the ice, the pressure being very heavy; breaking the floes in different directions; the ship apparently in great danger.

14 and 15 SEPTEMBER

Saw the land bearing as before, but at a greater distance....

22 SEPTEMBER

Latitude 74°08′N. Still close beset; sunk a small cask of powder in the wreck, and blew up part of her decks.

23 SEPTEMBER

Latitude 74°30′N, Longitude 63°05′W. About ten P.M. a very heavy pressure, the ice run over the top of the wreck. Our ship apparently in very great danger; every thing on deck in readiness, in case of any accident happening to our ship.

24 SEPTEMBER

Blew up part of the wreck with gunpowder, but still no provisions appeared; surrounded by a barrier of ice in every direction; and no water to be seen from the masthead.

25 SEPTEMBER

Latitude 74°16′N. This day weighed out every man three pounds of bread for seven days; no appearance of water....

27 SEPTEMBER

...Severe frost; no water to be seen in any direction. Appearances very alarming, and the people began preparing boats to leave the ship, in case of any accident happening to her.

28 SEPTEMBER

Latitude 74°14′N. Saw a little water making from N to SE; the East Land in sight, distant about fifty miles. Frost very severe, and our ship frozen in fast, and no possibility at this time of making any progress....

30 SEPTEMBER

Latitude 74°08′N. Water still in sight to the north and eastward, so that we were for several days tortured with a view of relief, but it was not in our power to reach it.

1 OCTOBER

...Water still continued to be seen; frost very severe; ice completely frozen together; every thing had a most awful appearance....

4 OCTOBER

Latitude 73°59′N. The ice still continuing in the same state, and the Dutchmen being fully convinced that there was no possibility of getting up their ship, and but little probability of obtaining provisions from her of any consequence; and there being every prospect of our ship remaining frozen up during the winter, and as our stock of provisions was totally inadequate to the support of both ships' companies, these unfortunate men, actuated by an independent spirit and from feelings highly honourable, held frequent consultations among themselves, how they might relieve us from the charge of their support, which we had no means of continuing much longer; and proposed to take their departure with boats over the ice, till they could get into some water; by which means they hoped to reach a Danish settlement, called Lively, a distance of about 340 miles. They soon communicated to us their intention, and the necessary arrangements were made for commencing their hazardous enterprise the next day. Thermometer 16° F below zero.

DUNCAN'S NARRATIVE PRESENTS the decision of the Dutch to leave the comparative security of the ship and make for the coast of Greenland as one arising simply from their qualities of independence and honour. A fact that Duncan tactfully refrained from mentioning, however, was put on record later by Captain Hoekstra in his own published account. On 25 September the Dutchmen had overheard the officers of the *Dundee* complaining to Captain Duncan that they were not getting enough food to live on because the ship's already scarce provisions were being shared with the wreck survivors. Duncan had pointed out that it would be "inhuman" to ask the Dutch to leave – they would surely perish. Nevertheless the possibility of the Dutch attempting to reach the distant coast was discussed openly a few days later, and on 4 October the English officers demanded that the Dutch leave. Again Duncan opposed them but Captain Hoekstra and his men, probably sensing serious trouble in the offing, made up their minds to depart, even though they saw "nothing but death" in the prospect.

According to Hoekstra, Captain Duncan (our detached and unemotional Englishman) wept bitterly at their decision.

---

5 OCTOBER

At four in the morning called all hands; weighed out four hundred and a half [pounds] of bread, and gave to the Dutchmen, also three casks of pork and a quantity of spirits, for each of their three boats; and ordered a party of our people to assist them in dragging their boats over the ice. They proceeded on the ice about six miles; where leaving the boats, they all returned to our ship, and rested for the night; early next morning they again proceeded, accompanied by our people to assist them, as before.

6 OCTOBER

The Dutchmen, assisted by some of our people, employed launching the boats over the ice, towards a lane of water to the eastward. About meridian our people left them to their hazardous undertaking.

7 OCTOBER

Still close beset; no expectation of getting liberated out of the ice. One of the *Cicero*'s men, John Thompson, [had] volunteered to go with the Dutchmen, but after launching over the ice with them all night, returned back in the morning. The boats still in sight from the masthead; every person on short allowance.

8 OCTOBER

Latitude 74°06′N. Still lying fast bound in the ice. About four P.M. had the satisfaction of seeing the Dutch people with the boats apparently under sail, bearing about south by west, from us at a distance about twelve miles; which gave us hopes they had found water.

Though the Dutchmen leaving us was their own proposal, and in which imperious necessity had compelled us to acquiesce; yet, when the hour of final separation arrived, the scene was truly distressing; and it was evident that each party felt deep and sincere regret, at the perils to which the other was likely to be exposed; after repeated shakes by the hand, tears ran down their manly cheeks, and I believe every individual was greatly affected by the parting.

From the time the Dutchmen came on board, our constant endeavours were used to get up the wreck, but finding that to be impossible, we tried various means, by firing down rockets and blowing up part of the decks with gunpowder, to get at some of the provisions; but in this we were also disappointed, not being able to obtain any thing of consequence, except a number of old blubber casks; but even these were, to us, of the utmost importance, by serving to increase our stock of fuel, which was so nearly exhausted, that we were obliged to use as firewood, all the masts, yards, spars, &c, that belonged to the wreck; but for which, long before the winter would have passed over, we must have been reduced to the dreadful necessity of burning our own masts, yards, and every thing combustible above the deck; and thus have become totally defenceless against the fury of the ice, the winds, or the waves. ...

9 OCTOBER

Latitude 74°N, longitude 62°30′W. Lying completely frozen up in the ice. No water to be seen. People employed covering the quarter deck over with sails, to keep the frost from penetrating through the decks; put the people on the scanty allowance at the rate of two pounds of bread for seven days, with a similar quantity of beef and pork. Thermometer 8° below zero.

10 OCTOBER

Frost very severe; making all possible preparations for the approaching winter....

Emma, Gipsy, *and* Undaunted *squeezed in the dense pack ice of Melville Bay in 1857. Sketch by William Barron. (Hull Museums)*

THE CREW OF THE *Dundee* could only speculate on the fate of the Dutch sailors who had departed for the Greenland coast, a tantalizing seventy miles away. Some doubtless envied their escape from the icy prison to whatever lay beyond. Others must have preferred to remain with the ice-bound vessel rather than trust their fortunes on dangerous crossings of ice and water with small boats in freezing temperatures.

On board the *Dundee*, as the days shortened, temperatures dropped, and new ice formed steadily over leads and ponds, all hopes of extricating the vessel faded away. Captain Duncan set the men to work preparing the ship for the inevitable trials of winter. The carpenter caulked hull seams already burst by cold. Efforts were made to keep the rudder safe by sawing the ice around it; and later, when ocean swell imparted some motion to the pack, it had to be unshipped completely. Rope fenders were constructed to cushion the blows of the surrounding floes. A toilet, nicely described by Duncan as "a small place of convenience," was built upon the ice.

The proximity of icebergs caused uneasiness. On 18 December Duncan wrote, "It was awful to behold the immense icebergs working away to the northeast from us, and not one drop of water to be seen; they were working themselves right through the middle of the ice." Constant vigilance had to be maintained.

Christmas passed—"a most dismal one"—and the New Year of 1827 came in with little cause for celebration. The *Dundee* had been locked in the ice for almost five months, during which time the pack ice had carried her southward at the scarcely perceptible rate of two miles a day. The crew might have found some satisfaction in the fact that they were now at the latitude of Disko Island, 300 miles closer to the open Atlantic than they had been in mid-August when the floes closed around them, but the piercing cold and depressing darkness of the polar night effectively dampened optimism and heightened dangers. The sun had dropped below the horizon during the first week of November and had not yet reappeared.

17 JANUARY 1827

The land still in sight. Ship drifting to the southward. The ice very quiet all round us.

This day we had the happiness of seeing the sun's rays just above the horizon, after having been deprived of the influence of that great luminary seventy-five days, during which we had very little daylight; the cold so intense, that it was scarcely possible to stay five minutes on deck without being frost-bitten. Our only prospect a vast expanse of snow and ice. Cut off from all the world, and doomed to pass the long and dreary months in gloomy solitude. When the boisterous winds had subsided into a calm, the death-like silence was continually broken by the despairing groans of hungry bears, or the equally dismal howlings of wolves and foxes....

20 JANUARY

Tremendous gales, with snow. Surrounded entirely by icebergs. Ice ranging about. Several small holes of water seen in different places, occasioned by the icebergs. Very severe weather. Several of our people slightly frost-bitten. Tremendous gales, the drift flying across the ice at a great rate....

24 JANUARY

Latitude 69°16′N, longitude 57°15′W. Got an observation, being the first since the fifth of November, when in latitude 73°12′N, longitude 62°44′W.

25 JANUARY

The south end of Disko, bearing S by E, distant sixty-three miles. No water seen from the masthead. The ice was at this time lying very quiet....

29 JANUARY

Saw several icebergs to the southward of us. About 6 P.M., the ice in which we were frozen broke from north to south of us for several miles; we thought it was occasioned by the ice driving against one of the bergs that were aground. The crack was about 200 yards from the ship, and very alarming. The land in sight. Ice ranging about. Every person on the alert.

30 JANUARY

Ice ranging about very much. Every person on the alert. Several icebergs round us. Carpenter caulking round the ship....

1 FEBRUARY

Several small holes of water a little distance from us, and in the water several whales. Called all hands. Coiled the boats' lines, and then launched two boats over the ice to the water.

2 FEBRUARY

At two P.M., John Barns got fast to a whale. At five killed her, towed her as near the ship as possible; laid her pass [sic] for flinching at daylight, alongside of the ice. All hands employed flinching and dragging the blubber across the ice on hatches. The bone 9 feet 4 inches.

3 FEBRUARY

At five P.M. Got the blubber off the fish, so knocked off until daylight. Eight A.M., called all hands, in order to take the whalebone out of the fish's head; but the weather being so very severe, we were obliged to come on board, most part of the people being frost-bitten.

4 FEBRUARY

...Called all hands, and went over to the fish, and got the whalebone on board. At meridian finished; hauled all the boats close alongside, and set the watch.

5 FEBRUARY

Latitude 68°40′N. Saw the land, distance from us about fifty miles. No water to be seen. Ice beginning to range about very much. The taking [of] this whale seemed to renovate the spirits of the crew, who killed the bears, foxes, and sharks that came to feed on the crang [carcass], and which afforded the men with the crang itself, many hearty meals for near six weeks, in addition to their limited allowance.

6 FEBRUARY

...Ice pressing very heavy close astern of the ship; we were much afraid of the ice giving way upon us; every person on the alert, for the preservation of the ship and their own lives. Latter part, ice slackened after working in and breaking our floe more than thirty feet.

7 FEBRUARY

Ice ranging about very much. Saw several whales among the bay ice, at a great distance, but no possibility of getting at them. One of my dogs attacked a large bear on the ice, and struggled with him till we came up with lances, and killed him.

8 FEBRUARY

Saw several whales, launched one boat over the ice after them; at dark returned and set the watch. Ice still ranging about. Carpenter employed caulking the ship's sides.

9 FEBRUARY

Called all hands, and got casks out of the hold for making off the blubber, when we found all the casks that had had water in them were frozen, and most part of them their heads had bursted. The same dog that attacked the bear on the seventh, also caught two blue foxes, without materially injuring them. We kept them alive more than three months.

10 FEBRUARY

All hands employed breaking out the hold, and putting the casks on the ice. Cooper employed preparing the casks, assisted by different people....

13 FEBRUARY

Done making off the blubber; filled thirteen casks, containing twenty-five butts. Cooper employed repairing the casks....

17 FEBRUARY

A lane of water broke out round our floe's edge, not far from the ship. Saw three or four whales. Launched three boats into the water after them, about three P.M. Michael Lee got fast to a whale. Shortly afterwards John Lander got fast to another whale; she ran him out nearly four lines, and the fore-ganger broke. Five P.M. killed Michael Lee's whale. Mooring her alongside in the ice until daylight. Bone 8 feet 1 inch.

Casks for blubber or oil were taken north as shakes, then assembled as needed by the cooper. Sketch from the Orray Taft *sketchbook, 1864-65. (Kendall Whaling Museum)*

WHALEMEN RARELY BOTHERED to measure the whales they killed. It was far more convenient to measure the length of the longest slab of baleen extracted, which bore a direct relationship to overall body length and, more importantly, to oil yield. A whale with five-foot baleen would probably provide about five tons of oil, while one with ten-foot bone would yield roughly thirteen tons. The eight-foot bone of the *Dundee*'s whale indicated that approximately nine tons of oil could be rendered out of its blubber.

The margins of whaling journals are decorated with drawings, or stamps — sometimes crude, at others remarkably polished — that indicate the whales and other animals captured, or harpooned but lost, by the whaling vessel. The journal pages reproduced below and on the next page show whaling stamps (and one walrus stamp) representing a variety of styles and periods.

Below left and right: Abram *of Hull,* 1839. *(Manitoba Archives)*

Top left and bottom left: Neptune, *1823.*
(Hull Central Library)

Middle left: Margaret, *1812. (Hull Central Library)*

Above: Laurel, *1828. (Hull Central Library)*

18 FEBRUARY

Employed flinching at the edge of the ice. People dragging the blubber on hatches. Six P.M., done flinching. Set the watch. Ice ranging about very much. The people fed heartily on the crang; also on the sharks, bears, and foxes, which we killed while they were feeding on the carcass.

19 FEBRUARY

Latitude 69°03N'. Made off the blubber. Fine weather for the season. Saw the land to the SE, distant about sixty miles.

20 FEBRUARY

...Finished making off blubber, and set the watch. Filled twelve casks (twenty seven butts). Carpenter employed caulking. Cooper repairing casks.

21 FEBRUARY

Saw several whales, but no possibility of getting at them for bay ice. Frost very severe. People employed sawing the ice round the ship, to ease her in her dock.

22 FEBRUARY

...Frost very severe. An immense iceberg setting very fast up towards the ship. Very little water seen in any direction. The east land in sight.

23 FEBRUARY

Latitude 68°37'N, longitude about 63°w. The dreadful apprehensions that assailed us yesterday by the near approach of the iceberg, were this day awfully verified. About three P.M., the iceberg came in contact with our floe, and in less than one minute it broke the ice; we were frozen in quite close to our ship; the floe was shivered to pieces for several miles, causing an explosion like an earthquake, or one hundred pieces of heavy ordnance fired at the same moment. The iceberg, with awful but majestic grandeur, (in height and dimensions resembling a vast mountain) came almost up to our stern, and everyone

expected it would have run over the ship: the consternation and alarm became general; our men ran some one way and some another, in order to get provisions, clothes, blankets, the boats, and other things, on the ice.

Considering there was no chance of escaping, I directed the people to endeavour to adopt such means as they thought most likely to preserve their lives for the present, in consequence of which they all left the ship, and went upon the ice, except the surgeon, the cook, two seamen that were frost-bitten, and myself. Fortunately I had determined that no circumstance or danger should induce me to leave the ship, and if she was lost to share the same fate; and it affords me the most heartfelt satisfaction to reflect, that my acting on this determination was the means of saving the ship, even after every hope was gone. I fear it will be difficult to convey to my readers who have never been on the frozen ocean, a correct conception of the situation of the iceberg and the ship. The iceberg, as has been before observed, came up very near to the stern of our ship; the intermediate space between the berg and the vessel was filled with masses of heavy ice; which, though they had been previously broke by the immense weight of the berg, were again formed into a compact body by its pressure. The berg was drifting at the rate of about four knots, and, by its force on the mass of ice, was pushing the ship before her, as it appeared to inevitable destruction. After being in this situation more than half an hour, I resolved on trying the effect of hoisting the jib and fore-topmast staysail; this, by great exertion, I effected almost without assistance, soon after which the sails filled, and enabled me to cast the ship, and thrust her between the berg and the broken floe of ice. Language is too poor to attempt a description of the feelings of those left on board the ship, on seeing the horrific iceberg rapidly glide by us.

Our deliverance (at least from present danger) was now complete; and, at about six, we contrived to fasten the ship to a floe of ice. The men that had left the ship to go on the ice with the boats, provisions, &c. were also exposed to great danger; the berg ran over one of our boats, and we never after saw a vestige of her; they also lost a three-hundred gallon cask of bread, (almost seven hundred-weight) one large barrel of pork, several of the peoples' clothes and blankets, and a considerable part of the cooper's and carpenter's tools with which they had been at work on the ice, and many empty casks.

The iceberg being between the men on the ice and the ship, hid her entirely from their view, and for more than a quarter of an hour they naturally considered the ship was lost, and that they should never more behold their commander, the surgeon, or their shipmates they had left on board. In this dreadful situation they held a consultation, in order to adopt some plan that might afford a chance of saving their lives; but, when it is considered that these intrepid fellows were almost destitute of provisions; that their clothes and blankets were very scanty, and the cold so intense, that (I really believe) the hardiest of the most hardy could not have survived in an open boat even a single night; it must appear not very likely that the consultation would be speedily concluded, how long it would have lasted is not easy to conjecture; but it was most unexpectedly interrupted by one of the men, who, probably, had never taken his eyes from the iceberg, exclaiming, with great ecstasy, "the ship, the ship." The rest of the men were seized with a kind of frenzy on beholding the jib and fore-topmast staysail. But they had no other means of expressing their universal joy than by three loud and hearty cheers, which resounded over the ice for miles, and, no doubt, frightened many bears from their dismal haunts.

The men soon found their way on board with whatever had been spared by the relentless iceberg; their congratulations were hearty, sincere, and unsophisticated. Mirth, and a little indulgence in festivity, was the order of the evening; and all were grateful to a Merciful Providence for such a miraculous deliverance. The congratulations of those who had braved the danger on board, were equally sincere and hearty; and it may be easily conceived, that the joining the ship by the crew, was not less pleasing to us than to themselves.

MARCH BROUGHT NO RELAXATION on board the *Dundee*. The situation remained perilous. On one day fifty icebergs were within eyesight. One of them ground its way noisily through the pack toward the ship with frightening force, and passed not twenty yards away. Dangers were continually present and the outlook always uncertain, but there were good reasons for a cautious, suppressed, optimism. The crew were yet in reasonable health; days were getting longer and warmer; and all the time the pack ice was carrying its captive whaleship farther and farther south, toward freedom. Holes of open water were becoming more numerous, and on the sixteenth whales surfaced in one of them, close enough to be pursued by two boats, although without success. By the end of March the ship had drifted almost eight hundred miles south from her extreme northerly position of the previous September.

On 17 April the influence of the open waters of the North Atlantic was at last felt. A strong swell reached in among the floes and shattered them into "ten thousand pieces." Frantically the crew shipped the rudder and set sail. After eight months and five days as a prisoner of the ice, the *Dundee* was once again under way, responding to the wind and lifting to the waves. High spirits reigned, and the captain promptly increased the weekly allowance to four pounds of bread per man with a little beef and pork. Within a week they began encountering whalers arriving for the 1827 season, and were able to barter gear salvaged from the wreck of the *Harlingen* for coal, fresh water, flour, bread, beef, pork, potatoes, turnips, peas, and barley.

Incredibly, Captain Duncan resolved to take up whaling again and continue through the season rather than return home directly, even though he observed that several of the men showed signs of scurvy and many others had little strength left. He ordered the boats after whales on the 25th and each of the two days following, but the crew disapproved. On the 27th

Duncan noted "several of the men very poorly this day, came aft, and requested to go home, as they were all very weak, and not able to begin another fishing." One suspects that the men's "request" was not presented in such timorous language. In any case, when pack ice closed around the ship a few days later and once again rendered her helpless, Duncan came to appreciate the good sense of the crew's sentiments, and decided to head homeward at the next opportunity. The whaleboats were brought on board for the last time on 7 May and the *Dundee* proceeded eastward, away from the pack ice and bound for Scotland.

Four weeks later the crew were rewarded by the happy sight of the Shetland Islands, Scotland's northern outliers. The *Dundee* sailed into Saint Magnus Bay on the west coast of Mainland, and dropped anchor in the harbour of Voe. Captain Duncan took the men ashore to church to thank God for their merciful deliverance. Certainly they had much to be thankful for. Aside from one death through natural causes in the previous May no lives had been lost on the voyage, despite the short rations enforced during the winter drift, the constant cold, and the ever-present threat of the ship's destruction by ice. Very few subsequent ice-drift voyages would be attended by such good fortune.

During their enforced winter in the pack ice the men of the *Dundee* must have wondered from time to time about the fate of the crew of the wrecked Dutch whaleship *Harlingen*, who had started for the Greenland coast in early October, dragging their boats over the pack. Their story emerged in 1828 in a book written by Captain Hoekstra and published in the city of Harlingen.

They had suffered unimaginable hardships in their six-day journey to the land. The boats were worn thin and holed by the abrasion of the ice, and had to be caulked and repaired several times; the men exhausted themselves by pulling the boats over hummocky ice floes, lowering them into open leads, rowing, breaking young ice, and hauling the boats up when leads closed; cold made sleep next to impossible, and extreme thirst assailed them at all times; hands and feet were frequently wet, three men fell into the sea between ice floes, and frostbite became common. When they were finally able to get into open water, they raised the masts and hoisted sails, but heavy seas kept them constantly bailing and spray froze all over the tossing boats and their frail occupants. But they made the land.

Ashore they managed to light a fire, using floorboards from the boats, and as a kettle of ice warmed over the flames the men, animal-like, fought each other to be the first to drink the water. One man, unable to walk by this time, died that day. The next day the remainder sailed and rowed southwards along the coast, pumping the leaky boats more or less continuously until they reached the settlement of Upernavik on 15 October.

The most dangerous part of the voyage was over. Help was at hand; the Danish trader Georg Jacobsen did everything possible for the emaciated, frostbitten men. After a few days rest they continued southward in the falling temperatures and declining light of descending winter, reaching Jacob's Bay at the end of October. They spent the winter in this region, dispersed among several small settlements so as to avoid overstraining the limited food resources available to the Greenlanders and the Danish traders, upon whom they were completely dependent. From the point of view of the Dutch the native huts in which they lived were crowded and uncomfortable, but there was plenty of food, water, and warmth. Only one more death occurred, from the lingering effects of severe frostbite.

When the *Dundee* finally dropped anchor in the Thames on 25 June the Dutch were still on the Greenland coast. Captain Hoekstra arrived at the home port of Harlingen on 24 August, accompanied by fourteen of the crew, including his own son and two other boy apprentices. Everyone had been presumed dead for some time. Other men arrived in subsequent weeks. A few, apparently undaunted by their dreadful experience, had shipped aboard British whalers on the Greenland coast at the beginning of the season, and did not return to Holland until mid-October.

Flensed carcasses, when they did not sink, often
washed up on the shore of Baffin Island.
(National Library of Scotland)

Three   The Dreary Coasts

The sun set as beautiful as ever I had seen it do in our own country. I stood for a time on the bows to watch the cirruscations of the aurora borealis, which were at times so bright as to pale the lustre of the moon, which then shone in full splendour. Surgeon of the *Hercules*, on reaching Greenland, April 1831

When we entered the mouth of the Sound the greenness of the hills was refreshing to our eyes after so long seeing the dreary coasts of the arctic regions. Surgeon of the *Hercules*, on returning to the Orkney Islands, October 1831

ACCOUNTS OF SHIPWRECKS, ice drift, starvation, frost bite and scurvy are common in the annals of arctic whaling. It was the spectacular narratives that tended to reach publication most readily (as they do today)—narratives describing voyages marked by disaster rather than success, by exciting events rather than humdrum routine. Because of this the image of constant crisis and frequent catastrophe persists in our mind's eye. Danger was certainly present, of course. Vigilance, determination, and courage were undoubtedly required in generous measure. But there were also times of relaxation and rest; bonds of friendship evolved among crew members; grog and good times punctuated the sailing and whaling routines; and the beauty of arctic seas and coasts could inspire even the most ordinary of seamen. There were times of boredom as well, especially for the sailors, whose range of diversions and freedom of action was more limited than that of the officers.

The man most capable of enjoying himself on the voyage was the ship's surgeon. His duties were minimal until accident or ill-health intervened. For much of the time he was a free agent, more or less able to design his own leisure activities on the vessel or on the nearby ice or land. As a medical man he was literate, usually had a lively interest in natural history, and was often eager to observe and record the unfamiliar phenomena of the arctic world. His gear normally included musket and fowling piece, and it was not unusual for him to undertake the collection of plants and animal skins or bones for shore-bound scientists. The ship's surgeon generally had the time, the ability, and the inclination to record his experiences systematically in a journal, often enlivening the account with sketches. To such men we owe a deal of gratitude; their daily entries provide much information about the nature of arctic whaling, including details that the whalemen themselves usually regarded as too mundane to be worthy of assigning to the pages of a copy book.

The surgeons of the whaling fleet were normally medical students rather than qualified doctors. They were attracted away from the classrooms, libraries, and laboratories of Edinburgh and other cities for a bit of adventure, a brief change from the academic life, and a not inconsiderable amount of money. The most famous of this breed was probably Arthur Conan Doyle, the creator of Sherlock Holmes, who sailed to the east Greenland fishery in 1880 on board the Peterhead whaler *Hope*. As he later wrote, "I went in the capacity of surgeon, but as I was only twenty years of age when I started, and as my knowledge of medicine was that of an average third year's student, I have often thought that it was as well that there was no very serious call upon my services."

If Conan Doyle's skill with scalpel and sutures was no better than his ability to keep his feet dry on the voyage, we can readily concur with these sentiments. One day, while seated on the ship's gunwales watching the men kill and skin harp seals on the pack ice, he fell overboard into the water. The captain, thinking he might be safer among the crew, invited him to join them on the ice, but he fell in twice more during the day. So impressive was his propensity for taking to the water that he was henceforth referred to as "the great northern diver."

For some medical students, however, there were serious emergencies, and these inexperienced young men often rose magnificently to the challenge. Charles Edward Smith, surgeon of the *Diana* during her ice-drift of 1866-67 in Davis Strait, was one. He assumed a large measure of leadership and responsibility after the death of the captain, and is generally credited with bringing the ship and many of her crew through safely.

The author of the narrative that follows does not identify himself, but by his knowledge of bird species, curiosity about all forms of wildlife, powers of observation, and attention to detail, he unconsciously reveals himself as the surgeon of a whaler, on his first voyage to the arctic regions. Neither is the vessel named in the journal, but circumstantial details in the text show that it must have been the *Hercules*, of Aberdeen, Captain J. Allan.

The *Hercules* made sail from Aberdeen in mid-March, 1831, in a rising southeast wind, which soon had the surgeon and a passenger prostrate in their bunks with seasickness, and which prevented the captain from stopping off Peterhead to discharge the stowaways—no fewer than six of them! The adventuresome boys were thus ensured a trip at least as far as the Shetland Islands. At Lerwick the officers went about the business of recruiting additional men and purchasing fresh produce. This was a common stopping-place for British arctic whalers and the traffic provided an important demand for local livestock and

poultry, and for labour, during the nineteenth century. It was perhaps a mixed blessing. Much in the same way as the Canadian fur trade drew young malcontents away from the drab seigneurial settlements of Quebec to pursue the vigorous life of *coureurs de bois*, thus depriving the farms of strong arms for planting and harvesting, the hiring of able-bodied Shetlanders and Orkneymen subtracted for half of each year a valuable portion of the work force of the islands.

On 20 March the ship weighed anchor and sailed away for Davis Strait. The surgeon, despite almost constant seasickness, somehow found the time and inclination to observe the forms of animal life encountered. Porpoises and seals were sighted from time to time, but he reserved most of his enthusiasm for the abundance and variety of bird life, including fulmars, kittiwakes, dovekies, murres, glaucous gulls, ivory gulls, snow buntings, ducks, and a titlark. Indeed a man interested in birds could hardly do better than to accompany an arctic whaleship, which was sure to attract a retinue of flying attendants, all eagerly anticipating the tasty food scraps and mouthwatering remains of flensed whales and other animals that would periodically be discharged into the greasy wake of the ship. Sea birds were a normal part of the voyage, and often a source of amusement to the sailors, who could easily catch them with hook and line, and then subject them to various indignities and

Left: A journal sketch, probably of the Hercules *under shortened sail in a strong gale, made by the ship's surgeon. (Aberdeen University Library)*

Seabirds of the eastern Arctic. Right and far right: thick-billed murres and common murres (looms); below, left to right: black-legged kittiwake; male common eider; and glaucous gull. (Barry Ranford)

cruelties. They liked to tie their wings and pair them off against each other like fighting cocks, or to release a soot-blackened bird into the air and watch it be picked to death by the white, colour-conscious members of its own race. There were doubtless other forms of diversion for the men as well. Some probably devoted their time to scrimshaw work—the carving and decoration of baleen and ivory—and others claimed to enjoy the somewhat esoteric hobby of making salt cellars from the sclerotica of whales' eyeballs.

Little more than a month out of Lerwick the comfortable routine of shipboard life was rudely interrupted by the death of a seaman from Shetland, "as fresh-looking a man as was on board the ship" and the only support for his widowed mother. All hands watched as the body was consigned to the spacious sea, but the coffin, although weighted with sand, would not sink, until at last someone managed to pierce it with a boathook, after which it filled and slowly disappeared into the depths.

On the first of May, as a garland floated from the fore topgallant mast, Neptune came on board the vessel to hold his court. With a good deal of ceremony he shaved all the green hands (the surgeon included) with a lather not guaranteed to give the smoothest shave—a mixture of "tar, grease, and all sorts of filth." Grog was then issued; good spirits reigned for awhile; a fight ensued later; and in the end the master had to restore peace by descending into the crew's quarters to "thrash among them with a stick."

They were by now at the edge of the whaling grounds. On 2 May an ice blink revealed the presence of pack ice ahead, and soon after they spoke the Dundee whaler *Advice*, a pair of whale jaw bones hanging over her taffrail to signify the first kill of the season. A week later the *Hercules* was beating up Greenland's west coast with thirty-four other ships in sight.

In hopes of reaching the rich whaling grounds of the West Side as soon as possible, whaling captains usually pressed northward along the Greenland coast as fast as the retreat of the ice would allow. By sailing, towing, and tracking, the ships clawed their way up the shore lead towards Melville Bay, one after the other, bowsprit to sternpost, their masters reckoning every additional yard an advantage gained. Progress was controlled by ice conditions, and when onshore winds closed the shore lead – their highway north – there was nothing to do but wait until it opened again. The frenetic activity associated with working the ship along the floe edge was suddenly replaced by comparative idleness. There were still routine tasks to be carried out on board ship, to be sure, but if the situation appeared stable the men could sport about on the ice, and perhaps indulge in some football or other games. If land was close at hand the captain would probably dispatch shore parties to take on fresh water, collect eggs, shoot birds, or hunt caribou, to provide a dietary change from salt meat and hard tack. Such excursions were a welcome diversion after a few months at sea. The ship's surgeon, who was usually free of shipboard duties, well armed, interested in plants and animals, and anxious to get a glimpse of the native Greenlanders, often accompanied these little expeditions. For the officers and men of the *Hercules* in 1831 there was an additional incentive to get ashore on the Greenland coast, namely to seek out survivors of a mutiny on board the Greenock whaler *John*.

In 1829 the *John* had been purchased to carry supplies for Sir John Ross's second expedition in search of the Northwest Passage, but at the last minute the crew of the whaler demanded guaranteed wages equivalent to a full ship of 200 tons of whale oil (even if they caught no whales at all). Ross refused to accede to the demand. His expedition sailed on the *Victory* without the *John* and spent four winters in the Canadian Arctic, losing only a few men, not far from the area where every man on the ships of Sir John Franklin was to perish less than two decades later. In retrospect it was probably very fortunate that the whaler *John* did not accompany the *Victory* as originally planned, for among her crew were some troublesome and dangerous men. In the following summer, during a whaling cruise in Davis Strait, they mutinied, apparently killed their captain, and set the mate,

surgeon, and a few others adrift in a whaleboat in Melville Bay.

After suffering great hardships the castaways managed to reach native settlements in the district of Upernavik and spent the winter of 1830 – 31 there. At home the men were presumed lost and Captain Allan of the *Hercules* was alert for evidence of their fate as he proceeded along the Greenland coast in 1831. As events turned out he succeeded in locating the ex-surgeon of the *John*, still very much alive, and was able to learn something of his experiences (see 15 June, below). Much later in the season, during inshore whaling on the distant coast of Baffin Island, the men of *Hercules* came upon the gaunt timbers of the *John* littering a beach. In an ironic twist of fate the ship had been wrecked on the rugged coast and many of the mutineers had been drowned.

During May, as the *Hercules* pressed northwards along the margin of the Davis Strait pack ice, the surgeon recorded new sights, sounds, and experiences in his journal. Observations and impressions crowd spontaneously one upon another in this unpublished account. He remarked upon the appearance of the floating ice, "some pure white shaded with sea green," and some "a dirty bottle green"; whenever the ship struck a large piece the shock "made her old ribs quiver and groan." On several occasions

The Hercules *sailing through scattered pack ice. (Aberdeen University Library)*

the boats pursued whales without success; one boat remained fast to a whale for seven hours only to have the harpoon draw out of the animal before it could be killed. Birds were plentiful, and in his spare time the surgeon shot at dovekies, fulmars, and ivory gulls. If he yearned for a larger game he got the chance on the twenty-fourth when a walrus was sighted near the vessel.

Butchering walrus on the Maud, *1889. Photo by Livingstone-Learmonth. (Public Archives Canada, C-88291)*

Left: Cutting out walrus tusks. Ivory was a valuable commodity in the late nineteenth century. Photo by Livingstone-Learmonth (Public Archives Canada, C-88292)

Bottom left: Scraping walrus skins. Photo by F. N. Gillies. (Dundee Museums)

TUESDAY 24 MAY

This morning I was roused to go after a sea horse [walrus] which lay upon a piece of ice near our vessel. When we approached the huge beast he lifted his head from the ice, looked at us, and laid it down again. When we approached within twenty yards we fired two balls into him and he was unable to take the water for some time. We thought all was right and did not hurry ourselves to come up to him, but we got our lances ready which gave him time to recover and slip into the water. He rose a short distance off bleeding so as to tinge the sea for some distance round; however, he dived. We were obliged to go on board as the vessel was going before the wind. He was afterwards picked up by the *Comet* of Hull.

Soon after we went after a seal which had a Yack's [Eskimo's] drag fastened to him. The drag showed us where he swam. I fired several balls at him and put two into his back. Candie dispatched him with a harpoon. When he struck a great quantity of blood and air issued from the wound. The Yack's harpoon had forty-two feet of line attached to it. The seal had been struck several days before. The drag was a small sealskin inflated. The line was made of sea-horse skin. We had a hard pull to reach the vessel. The seal measured [five?] feet six inches. Its blubber filled a large tub. It was a male. We saw numbers of white whales and sea unicorns [narwhals].

SATURDAY 28 MAY

...On Friday we made Barry's Island, the southwest of the Frow Islands, lat. 73°N. We made fast to a low berg under lee of Barry's Island. Six other ships also made fast to it. A boat's crew of the *James* came on board which was lost to the westward among floes. In the evening Captain Allan and I went ashore, landed on Barry's Island. Numbers of ducks were flying among the islets. I knocked down a drake but he dived and we never saw him again. Soon after R. Crombie killed a burgomaster. We wounded a small seal but lost it.

SUNDAY 29 MAY

Calm. Still made fast to the berg. We saw eleven Eskimo go on board two vessels about two miles to leeward. Half past eleven P.M. set out in search of one of the men of the *John* of Greenock who was said to be living among the Yacks and who had lost his feet from the cold, and being thus disabled was obliged to remain at Upernavik. We pulled close to the land in about two hours, the ice being pretty open. We found on approaching the land that there was a barrier of ice which opposed our landing but as it was moving to the south we landed on a rocky islet and walked about to keep ourselves warm till an opening should present itself. About two hours after, the ice slacked and — an opening presenting itself — we pushed in. In a clear hole of water close to the rocks we fell in with a flock of dovekies. I wounded one which was still able to dive. The water was so very clear that we could see it below the water. The water was so clear that we could see the smallest object at a considerable depth. The bottom is covered with two species of fuci [seaweed]... The water of the Arctic Ocean teems with animal life even more than that of the seas of temperate climates. I saw numerous medusae [jellyfish], some of them very beautiful but so fragile as not to bear handling. They had a gelatinous body with beautiful red, feathery arms proceeding from it — *Clio borealis*? [In fact these are not jellyfish but pteropod mollusks known as "sea butterflies."] They form the principal part of the food of the whale.

We landed upon a small, rocky island which like most of the rocks of this country was of a red colour and very brittle and dangerous to climb among. They were covered by a species of black lichen. I saw withered grass upon the rocks seven or eight inches long, and crowberry and heathberry plants. We crossed from this island to another much higher in order to look out for the Yack's huts. We commanded a very extensive prospect from the top. We saw the land split into innumerable creeks, promontories and islands as far as we could see. We could see no continuous land. The air was so pure and calm that we could converse with our comrades about a quarter-mile off. I heard the song of the snow bunting, which sounded very sweet among the bare rocks. We pulled into creeks and round islands in all directions but could not see the Yacks. I killed an eider duck and a dovekie. We saw immense numbers of eider and king [eider] ducks in pairs and flocks. We saw some very grand prospects among the rocks, which were in part still covered with snow.

About noon the sunlight was very strong, the water running from the bergs. The water was calm as a mirror, reflecting the rocks and bergs in a beautiful manner. We lay down in the boat and slept for some time, after which we pulled northwards, pushing the boat occasionally through heavy streams of ice. We pulled into a fiord and at length saw a hut at the foot of a high rock. We pulled up to it but found it had been deserted for some months. We saw the tracks of foxes about the house. It was built of stones and roofed with wood and turf. The doors and windows were made of seal's skin but were torn away by the dogs and foxes. We found the floor thickly covered with seals', unicorns' and dogs' bones, and pieces of leather and fur, which gave out a most offensive smell. We fell in with some Yack graves which were situated at a short distance from the hut. They were made like a drain, being formed of stones and roofed over with flat stones. We pulled off the roof of the tomb and found that four bodies had been laid in that grave, some of them for a very long time, as the bones were quite visible. They had been buried with their clothes on as one of them had a sealskin cap on his head. Their crania presented some very distinguishing marks, especially their great breadth across the cheek bones. The superior maxillary bones were very prominent, especially on a small skull which I supposed was that of a female. I wished very much to have kept one of their crania but the boat's crew would not hear of it.

We left the island and returned by the way we came as ice blocked up our passage to the westward. We passed a very

high cliff where the burgomasters appeared to have fixed their breeding place, as they kept flying along its face. It was so high that they appeared like flies. We saw several cormorants flying up the fiord. When we entered the ice it was pretty open. We saw a burgomaster tearing at something in the water. On pulling up we found a small dead seal. It was quite fresh. The burgy had torn out its eyes and opened its skull and taken out the brains, which was more easily affected as the seal was very young and its skull soft. The burgomaster has been seen to pursue the rotchies. The seal's skin was beautifully spotted and ringed with black on a silver grey ground. It was about three feet along.

When we approached within four or five miles of the ships we saw that the ice had come down from the north. There was a continuous floe of many miles square which coming down pushed all the smaller pieces together so as to close up the lanes. We began to launch the boat across the ice, which we found was excessively fatiguing. We continued at the laborious task for several hours till we were completely exhausted and we saw that we had scarcely made a quarter of a mile nearer the ships. Our provisions were quite exhausted. We had thrown away the seal to lighten the boat. We lay for five or six hours and suffered greatly from the cold as we were all wet through from falling into the holes of the ice in launching the boat. At length to our great joy we saw some men travelling towards us over the ice. They hoisted a jack as a signal, which we answered. They reached us in about an hour and a half bringing provisions and rum, which relieved us greatly. In about five hours after, we reached the ship very much exhausted. They scarcely expected ever to see us again.

After having slept for four hours I was wakened to see some Yacks which had come on board. Soon after one of them entered our cabin. He was a little, stout-made fellow about five feet five inches high, about twenty-two years of age, of a clear olive-brown complexion, with long coarse black hair hanging about his ears. His face was broad and his cheeks blown up with fat like bladders. He had on a

*Cape York Eskimos photographed by
Livingstone-Learmonth in 1889. (Public
Archives Canada, C-88395)*

white cotton sort of jerkin with sealskin shirt, sealskin breeches and boots, with a white sealskin cap on his head. He offered a sealskin jerkin for sale. I offered him a pair of cloth trousers for his jerkin, which after a minute inspection he took. He had white whales' teeth for buttons on his jacket.

The old Yacks had very dark mahogany complexions. Their eyes are placed far apart. One of the Yacks got rather tipsy. His brother took the greatest possible care of him, following him everywhere lest he should fall overboard. Our sailors and the Yacks began to dance between deck. Some of the Yacks appeared to have good ears as they kept time to our music pretty well. Their dancing was very clumsy. Most of them have sufficient English to enable them to buy and sell. Some of the Yacks got rare bellyfulls of blubber and whaleskin on board the *Traveller*. One old fellow devoured about three pounds; he kept always saying *very good*. They slept all night on board our ship as the ice was not in a good state for getting to the land. I put my double percussion fowling piece into one of their hands. He examined it very minutely, especially about the lock, and returned it saying that it was "no good." I began to fire off caps at a candle. The Yack showed that they were well used to shooting. Their guns are all rifles and they are said to be deadly marksmen.

I have often heard the sailors speak of the wild Yacks of the East Side under the name of "red-eyed Yacks." I am told that they come down from the interior and plunder the Danish settlements occasionally.

TUESDAY 31 MAY

We are now made fast to a berg about thirty miles to the northward of our former station. In reaching in to make fast we touched a rock on which the *Mademoiselle* of Dieppe, which was following us close, grounded before she could be warned of it. In the afternoon she was pulled off by the *Comet* of Hull....

FRIDAY 3 JUNE

Plying to windward. Horse Head bearing NE ten miles.... At eight o'clock A.M. all hands were called and the boats lowered away after fish. The wrecked harpooner struck one which died on the third harpoon. It was killed about seven or eight miles from the ship so I did not see it. We were standing away from the boats towards the land when the jack was hoisted. We tacked and stood towards them. The fish was towed alongside, the sea breaking over it like a sunken rock. Little of it was to be seen above water but the two fins which were tied together across the breast. The skin of the whale offers very little resistance as the fibres go from without inwards and the epidermis is very thin. I could easily push my finger through the true skin after scratching through the epidermis. The colour of the skin is deep black. The under jaw was pure white and the belly mottled with white. When the gum and bone was hoisted on deck it had the appearance of a huge black hairy curtain. She was about forty-five feet long, nine feet bone, and filled twenty-five butts or ten tons of blubber. The blubber was about from twelve to fourteen inches thick.

Left: Sketch of Greenland woman from a journal on the Diana, 1863. Reproduced by permission of Dr. Peter Wells, Cheshire, England, from the original logbook in his possession. (Hull Museums)

Baleen coming on board. Photo by Livingstone-Learmonth. (Public Archives Canada, C-212)

TUESDAY 7 JUNE

Today I saw great numbers of white whales. Their vision is very acute. They never rise very close to the ship. I never heard of their being struck by the whalers. The Yacks kill them with the harpoon and drag. The *James* of Peterhead found one dead off Hare Island with the harpoon fixed in it; the drag had burst but was still attached. It produced about half a ton of blubber. I had often heard from old fishermen that they whistled when deep under water. I had an unexpected opportunity of verifying this observation. Great number of whales were to be seen round the vessel. I could see them passing under the ship's bottom, their whiteness making them distinguishable at a great depth. I was standing in one of the chain boats with my gun loaded with ball ready to fire if any of them should rise near the vessels. One of the men who was standing by me remarked that he heard the white whales which were passing under the vessel whistling. I listened and could hear them distinctly making at times a whistling, at other times a chirping noise which varied in intensity according to the depth at which they were situated.

When the *James* of Peterhead was lost about fifty miles to the westward of the Frow Islands the ship's company, when they had their chest and bedding on the ice, were much alarmed by the boldness of the bears, which were very numerous and savage with hunger. Some of them tore open their chests and scattered their bedding about. One bear was seen to make a spring off the ice at a living whale which was lying close to the edge, so that they were obliged to back off and were prevented at that time from striking in a second harpoon.

THURSDAY 9 JUNE

This morning I was called out of bed as there was a bear on the floe edge. I got out as quick as possible. I saw the animal standing about 100 yards from the ship. Before I could get loaded one of the men fired a ball at him. He turned over heels uppermost but immediately got to his feet, gave a loud

growl, and walked away towards the land. I and two others got upon the ice and pursued him. When he saw us he began to run and soon distanced us. We saw drops of blood at every footstep. We followed him about two miles. He was rather a small animal; his footsteps were about [four?] inches across. When close to the ship he reared himself on end and sniffed the air.

SATURDAY 11 JUNE

Calm. Loom Head bearing E, distant fifteen miles. A boat was sent away after a large sea horse which was in the water. After pulling after him for some time we almost got within striking distance of him, but unluckily he observed us and dived. In the afternoon we went among the islands. The ducks had just begun to build their nests and deposit their down in them. R. Crombie and I travelled from one island to the other on the ice. I killed three sandpipers on one of the little rocky islets *(Tringa maritima)*. We saw many eider ducks, burgomasters and long-tailed ducks [old-squaws]. They are called by the singular name of "coal and candlewicks" from their cry, which is very musical. I heard

Left: Bear cub and dead mother.
(Dundee Museums)

Below: Scraping a polar bear skin (Dundee Museums)

several loud reports like that of great guns, which I at first thought were fired from our ship as a signal for our recall. These thundering sounds arose from the falling of icebergs. I saw a berg upset. It caused a heavy sea all around it for the distance of a quarter of a mile so as to make our ships roll at their moorings. Before it upset it was much higher than our masthead. It rolled about in the water a long time before it settled. We went on board the *Earl Percy* of Kirkcaldy, which towed us close to our own ship. Our ship was made fast to a grounded berg about a mile from the large Duck Island. The ducks are all scared from the islands so that very few are to be seen on them.

WEDNESDAY 15 JUNE

I saw the surgeon of the *John* of Greenock, which was lost last year. He described the manners of the Eskimo to each other as most admirable in all the relations of life. During all the seven months that he spent among them he never saw a son refuse to obey his father's bidding. There are 500 Eskimo belonging to the settlements of Upernavik. They are very hospitable and gave the wrecked seamen a share of what they had. The winter was very bad one for the sealing, as the continued gales of southwest wind never suffered the bay ice to form. The Eskimo at a settlement about fifty miles from Upernavik were obliged to eat the leather off their canoes and that of their boots, and had it not been for some assistance afforded them by the Danes they must have all perished.

He said that they take numbers of white whales, the skin and blubber of which, when frozen, they hold to be a great delicacy. They kill dogs only on great occasions such as marriages. They sometimes shoot the white whales with their rifles. The Eskimo fears to encounter no animal but the sea-horse, which sometimes sinks their canoes with his tremendous tusks. However, they kill considerable number of them. He once saw a boy of ten years of age kill an old bear and two cubs with his father's rifle, his father being asleep. They occasionally kill an immensely large species

of seal which they call ursuk [bearded] seals which, however, are very rare. These large seals are sometimes twenty-two feet long. The Danes charge them exorbitant prices for their articles in barter.

FRIDAY 17 JUNE

He said that the stories about red-eyed or wild Yacks of the East Side are quite void of truth. When an Eskimo attacks a bear he looses three dogs upon him. These distract his attention. He rises on end. The Eskimo whistles off his dogs and shoots the bear with his rifle. The Eskimo are a short-lived people. They do not in general attain a greater age than fifty or sixty. Eskimo infants are quite fair at their birth....

THURSDAY 23 JUNE

Lying made fast to a berg close to the northernmost of Baffin's Islands. We go daily on shore to shoot. We kill very few ducks as they are quite wild, being continually scared by boats from the ships....

SATURDAY 25 JUNE

Plying to windward with sixty sail in company. We have now entered the dreaded Melville Bay. The floe pieces which we pass bear marks of the tremendous pressure they have suffered. I have seen squeezed-up pieces on a flat-topped berg about forty feet high above the water. Grounded bergs detain the floes and hinder them from floating down the straits. Bergs, when carried by the current, help greatly to break up the floes. I have often seen lanes formed by the passage of bergs through the floes. The floes which destroyed the ships in Melville Bay sometimes went at two and a half miles an hour.

WEDNESDAY 29 JUNE

...There are many stupendous bergs around us, some of them full of rents and fissures which split them from their summits to the water. They appear quite ready to fall.

Accidents from the falling and upsetting of bergs are by no means uncommon. Their equilibrium is sometimes so delicate that the slightest force upsets them. They come down with a noise like thunder and make a heavy swell around for a great distance. Some of them form arches so that a boat could sail through them. Others have pinnacles like old ruined towers and various fantastic shapes. I once entered a cavern which reminded me much of the inside of a gothic cathedral. Its inside roof was bristled with icicles. The sides and roof were beautifully tinted with white and blue. It went a considerable way into the body of the berg. The bergs which have upset are easily known by their corners and points being rounded and polished by the action of the sea. When I viewed these immense bodies of ice I could not but smile at the vain theorists who attempt to account for their formation by the condensation and freezing of the moisture of the atmosphere. These people have never seen an iceberg. I have seen strata and veins of earth in them. Nothing is more common than to see them with one side quite blackened with earthy matter.

When passing close under a high, dangerous-looking berg it gave a crack followed by a sound as if the berg was rending asunder. We pulled away from it as fast as we could. We saw a very large seal get upon a small piece of ice. He had some difficulty in getting on it owing to its smallness. He rolled about with his paws in the air. We fired two balls at him as he lay on the ice. He lay and looked at us for a moment or two as if amazed, and then slipped into the water. We often go shooting along the floe edge. The rotchies [dovekies] are to be met with in immense flocks among the floes. I killed two snow birds [ivory gulls], a tern, and two rotchies. The rotchies when wounded always, if able, get upon a piece. They are sometimes so tame that they can be killed by throwing a stick among them. ...

TUESDAY 5 JULY

Four o'clock P.M. all hands were called to track the ship. Each man has a broad canvas track belt which goes over his shoulder like a sword belt. This they attach to the track

William Barron's sketch of whalemen in Melville Bay tracking ships northward in the shore lead parallel to the Greenland coast. (Hull Museums)

line. They can go four miles an hour with ease if the ice be good. It is very laborious work. Our men got a dram every three hours. After towing about two hours we were stopped by the closing of the ice ahead. We made fast to the ice...[and later] began to saw docks into the ice. They are generally two ships' lengths long.

WEDNESDAY 6 JULY

Nine o'clock A.M. Towing with about thirty sail in company. It is a very lively scene to see, the ships all so close together, to hear the songs of the men while tracking, pipes playing, the skippers [howling?] from the mastheads. The weather for the last few days has been beautifully serene. The sun is powerful during the day but it is very chilly at night, the sun being low in the horizon. At night the sky has the appearance of an English sky towards evening. A film of ice forms on the sea in open holes at night. The Devil's Thumb bears ENE distant twenty miles. I have several times of late seen what the sailors call a fog scaffer. It is like a rainbow without the prismatic colours. It is seen in foggy weather before the clearing of the air opposite to the sun. One of

our Shetland men had his leg seriously bruised by the breaking of a warp [rope].

THURSDAY 7 JULY

...On Thursday at twelve noon I was awakened with the joyful news that we had got clear of the ice and we were then running with all possible sail – ring-tails, skysails, boats' sails set, everything that would draw. Wind SE. Some vessels behind us are beset among the floes. We ran all day, thirty sail in company.

P RESENTED AT LAST with the long-awaited opportunity to leave the Greenland coast, the vanguard of the fleet made haste to get westward. Exhilaration faded on the following day, however, when no wind arose to move the ships and the sailors had to resort to the frustrating expediency of towing the ships behind the whaleboats. By manning the oars like galley slaves for almost eight hours the boat crews of the *Hercules* succeeded in hauling their vessel past many of the others, until only two remained farther ahead. (The surgeon took no part in these arduous and mindless labours, but amused himself by shooting seventy dovekies during the day, thirteen of them with one load of shot). On the next day the wind revived, but blowing from the south it drove the departed floes of the Middle Pack back into the upper reaches of Baffin Bay, imposing once again a substantial barrier of drifting ice. "We kept on running," the surgeon recorded, "thinking to push through," but this course of action proved futile, and by 9 July the *Hercules* lay surrounded by pack ice in violent motion, with the visibility obscured by rain, and a number of icebergs waiting ominously to leeward like reefs about to claim a disabled ship. It was a confusing situation, intensely dangerous, and eerily highlighted by the sinister sound of grinding ice and the plaintive wailing of seabirds.

The crisis passed. By 11 July the pack had opened enough to enable the *Hercules* and three nearby ships to proceed. Manoeuvring square-rigged vessels among drift ice, considering the difficulties of stopping their forward momentum in constricted waters, was a hazardous undertaking, which demanded a nice sense of judgement on the part of the captain, and instant responses by the officers and men. That Captain Allan was a bold—and possibly a skilful—ship-handler we may infer from the surgeon's observation that "we occasionally run before the wind at the rate of seven knots through openings between floes which scarcely admit the vessel." A week of this risky and exhausting work brought the whalers within sight of the West Land.

on board. That night the skipper got particularly [illegible word]. He insisted that the cook's duck should be brought to him that he might decapitate it with the carving knife. This gave great offense to the men as the duck was a great favourite with all hands. They hid it in a chest but the skipper got so outrageous calling for the duck that the duck was produced. Rather than see it killed one of the men threw it overboard. This put the skipper into a rage. He made a boat be lowered down to pick up the duck to have it killed. The poor duck followed the boat. It was then killed.

When the boats reached the land floe many bears were seen moving about, all of which set out for the land at full speed except one which lay down at the back of a hummock. The mate, R. Crombie, and a Shetlandman got between him and the land and drove him towards to the water. He turned once or twice to look at them and seemed inclined to give battle. However, they shouted and forced him to the floe edge, from whence he made a spring into the water. The two boats pulled after him and soon came up with him. When he saw that he could not escape he turned and charged the boat open-mouthed, blaring out hideously, showing his white tusks. He received a lance thrust in the neck from the boat's bow and the boat shot past, having a deal of way; the boatsteerer as he passed him pushed a lance into his side, which killed him. It was a male not very large, but had a remarkably large head. In his stomach was about a gallon of oil. His canine teeth measured two inches in length.

SUNDAY 17 JULY

This morning the West Land was to be seen high and bold about thirty miles off. We ran across the mouth of Jones Sound, the wind blowing hard from the NE. The sea was very rough in the mouth of the sound. In the afternoon the wind fell away. The land being about twenty-five miles distant, four boats were sent away which reached the land ice in six and a half hours. They were often pretty near to fish but did not get fast. There were about fifteen of us left

MONDAY 18 JULY

Running to the southward along the land. The land next the sea is smooth and low. Farther inland it is high and mountainous. There is snow in the channels and fissures of the sides of the rocks, and in places not exposed to the sun. Many fish running to the southward. One small fish rose seven or eight times parallel to the vessel, which was then going at the rate of six knots. It soon went out of sight ahead. We saw the *Dee* with her jack up. To the northward

Harpooning a Whale

of the north horn of Lancaster Sound I had the pleasure of seeing three glaciers. ... Their edges looked like a wall of ice, as I guessed about eighty or ninety feet high. This wall is shattered and rent by the fall of avalanches caused by the action of the sea on the base of the wall and by the torrents of water in the end of the year. At the top of the glacier the ice is light and porous but at the bottom it is more dense owing to the infiltration and subsequent congelation of the rain water. Thus are the icebergs formed which have caused so many disputes about their formation. The rocks projected through the glacier in many places.

Almost calm. All hands were called down after Diverty struck a fish which took out seven lines and had four harpoons struck. When a fish is lying, little is seen but the two eminences of the rump and crown. She can go away in three modes. When going on end [i.e., forward] they generally "make a back" as it is called; they lift their back out of the water. When lying they make backs when they move their fins. In going "tail up" they elevate their tremendous tail high in the air, when it is seen for a second or two; they go away in this manner when feeding. They

then go to a great depth. The other manner is going down "tail foremost." This they do when scared or struck. They then strike forwards with their fins like backing a boat. We saw a small sucker together with its mother. The sucker was very small, being only eight or nine feet long. We saw many narwals. Some of them are beautifully spotted like what are called coach dogs. We saw several young ones in company with their mothers; they were only two or three feet long. The young *white* whales are dark lead colour. The blowing of a whale is a prolonged sighing noise which can be heard in calm weather two miles off [and] continues about seven seconds when the fish is lying still.

FRIDAY 22 JULY

I went away in the boats with our specsioneer [chief harpooner]. The whales were extremely numerous. We could see seven or eight at a time. Both sides of the sound are visible. In an hour four of our boats had their jacks up [indicating they were fast to whales].

One [whale] was lost. They were all under size. The fish when dying blows pure blood so as often to dye the boats and crew red. They blow thick and roll about, lash the water with their tail. The sea is crimsoned with their blood

Far left: An American whaleboat under sail with the harpooner about to strike. One of a series of watercolour sketches by Timothy C. Packard, master of the Andrews *of New Bedford on voyages to the eastern Arctic in 1865 and 1867. The ship was wrecked near Blacklead Island, Cumberland Sound, in the autumn of 1867. (Houghton Library)*

Left: After being harpooned and lanced, a whale capsizes a boat, turns "flukes up," and dives. The boat on the right indicates with its jack that it is fast to the whale, while other boats approach to add their harpoons. Sketch by William Barron. (Hull Museums)

Top right: Lancing a harpooned whale. Sketch by Timothy Packard. (Houghton Library)

Bottom right: The laborious work of towing a whale back to the ship for flensing. Sketch by Timothy Packard. (Houghton Library)

for some distance round. The arm of the lance is pushed in about four feet into the animal. When accidents occur with fish it is more owing to the foolhardiness of harpooners than any evil intention of the fish. Four of our fish died on the first harpoon. Having gone to the bottom they had to be pulled up by heaving the capstan or by men walking the ice. Three were secured. The harpoon drew [out] of another.

SUNDAY 24 JULY

Today three fish were killed and one flinched, after which all hands went to bed....

MONDAY MORNING 25 JULY

Three falls were called this morning. Each boat's fish died on the first harpoon and all under the ice. We got out two but the harpoon which was in the other one drew.

TUESDAY 26 JULY

Busy making off. The ship is in a very uncomfortable state. Everything is out of order and the decks lumbered with casks, shakes, blubber, bone, &c. We have got twelve fish in the sound, three under size. There are four fish rafted alongside. Those which were killed on Sunday have swelled enormously. They have incisions made in them to ease them, as it is called when a boat hook or lance is pushed into them. The air which has been generated rushes out with great force and a loud noise. The intestines, part of the liver, and stomach have protruded, the walls of the abdomen having given way. The stench is very offensive. The mallemucks [fulmars] are excessively numerous about the ship. Their boldness is excessive. They may be seen standing on the fish and trying to tear through the skin but they can only get through the epidermis. They are so regardless of danger that they can be knocked down with a stick. I have observed a mallemuck dive to the depth of two or three feet after a piece of blubber or flesh which is sinking. They do this but seldom.

We are close to the south side of Lancaster Sound. The land is high inland and much covered with snow. The fish are all running up the sound. The sound is much frequented by the young fish. At the other parts of the West Side, Middle Ice, and Southwest fishing, the fish are all large. Whales entirely white have been killed. I have heard of three instances, two at the west side of the straits, the other at Greenland. I saw some of one, their whalebone. It was of a fine cream colour. Some of them have a deal of white on their skin—their under jaw, the eyelids, the under part of the root of the tail. Young fish are called "blue skins." The blubber of these is soft and very white. That of the old ones is of a deep orange colour. When a fish lies for some time in the water and begins to decompose the bone often falls out....

TUESDAY 9 AUGUST

$72°36'$N. Standing across to the Middle Ice. Blowing a stiff breeze. We saw many carcases of whales floating, hundreds of mallemucks and burgomasters on and about them. The sun almost dips at midnight, at which time the air is cold and chilly. We have many snow showers.

SATURDAY 13 AUGUST

Reaching off to the Middle Ice again. We were within thirty miles of the land on Thursday. We have been becalmed for the last three days with a few light airs occasionally. It is as dark at midnight now as in Britain about the 18 June. We are all in a high raffle owing to the skipper's bad conduct. He swears he will find fish by having plenty of old logs to consult. The land which we saw was in $70°28'$N—about the mouth of Clyde Inlet or Agnes Monument.

Right: Ships stand off the coast as their whale-boats cruise closer inshore, "rocknosing" for whales in the autumn. Sketch by David Cardno. (Mr. and Mrs. William Cruden)

For the last ten days we have had constant thick weather with heavy gales of wind. We have run to the northward to get round the tail of the ice which hinders us from getting in with the land. ...

ALTHOUGH THE *Hercules* sailed briefly northwards to avoid a barrier of pack ice her general course was towards the southeast along the coast of Baffin Island. This whaling late in the season was termed "rocknosing." As the ship anchored in a protected bay or stood off the land, the whaleboats would cruise close inshore to intercept whales migrating south to their wintering zone between Labrador and Greenland. For the whalemen it was hard, cold, uncomfortable work and, judging from the following account, it was not always rewarded by great success in the whaling.

Four boats were sent inshore. We had beef, bread, coals, and top gallant stay sails to make a tent if we stopped all night. We pulled in but could not reach the land for loose ice. ...

While writing the above one of the boys called down the companion "Doctor do you want to go after a bear?" I seized my gun and ran on deck, got into a boat which was lowering away. There were three bears on some...ice between us and the land, which was about ten miles off. We pulled among the loose ice to cut off the bears from the land. After pulling hard for about an hour we got between them and the land. We shouted, blew our fog horn, and made every possible noise to drive them towards the sea. They did not proceed very quickly as they had many lanes of water to swim over. We got pretty close, when [we] could see their black noses and hear them growling and blowing. They all got on a piece [of ice] in a body. I fired and wounded one of the small ones, however they all set off at a long gallop across the ice and took the water. We pulled after them with all speed. I fired at the old one but missed. They all swam close together; the mother appeared to caress her cubs with her nose. When we approached within about thirty yards of them the mother turned and charged us open mouthed. Candy our harpooner pushed his lance into her neck. She took the lance in her teeth and dived and came up roaring among our oars. She was so near that I could have laid my hand on her. She seized an oar in her teeth and crushed it. I thrust at her with my lance but could not get it entered right at the first push. After this she turned on her side and Milford finished her. She dropped her head into the water, her last look being directed towards her cubs who were blowing out and swimming out to sea. Robert Crombie shot one of the young ones with a bullet. The other kept swimming around his dead brother. Milford threw a noose over the other's head and drew him up to the boat's bow, where he hung [roaring?] and biting the boat's stem. He was strangled before reaching the ship. We towed the whole three to the ship and hoisted them on deck. The mother was rather small, only measuring eight feet seven inches in length and three feet high on the fore shoulder. The young ones were nearly as large as their mother, being six feet in length. They were very fat. Their stomachs were full of whale oil. The moon rose over the West Land tonight. The air very calm and serene.

TUESDAY 23 AUGUST

Today the boats were sent inshore. The land was about six miles off. The sandy beach was covered with whales' bones. The men lighted their fires and boiled their coffee kettles. They got about ten hundredweight fine whalebone out of some dead whales lying half buried in the sand. The blubber of one of the whales had a singular look. It was dried, the oil having [run?] out of it leaving the stringy tendonous fibres which had been pulled out by the birds so as to look like bear skin. There were many yellow flowers on the land, birds singing. I killed a raven and a sandpiper (*Tringa maritima*), of which last birds there were numerous flocks. Saw tern, ducks, burgomasters, sandpipers, in abundance. They found two dead seals which had the muscular parts removed by the insects, the skin remaining entire, the insects having entered by the mouth, eyes and anus. The sand was composed of white quartz with a few black particles of mica....

SUNDAY 28 AUGUST

Fine, clear weather. Reaching inshore, the *Rambler*, *Duncombe* and *Wellington* in company. We were told that the wreck of the *John* of Greenock was lying on the low shore. We called all hands and sent four boats ashore in search of a spar to replace our broken bowsprit. We reached the shore in about half an hour. It was a flat, sandy beach, the country for a considerable distance inland being flat, covered with grass and moss, with numerous yellow flowers. Numerous runs of water cross the low land from the hills. There was snow remaining in the dry beds of the torrents and in the ravines. There were mischievous sandpipers running about the water edge and large flocks of ducks in the sea. There were many whales' bones lying on the sands. We at length found the foremast of the *John* half buried in the sand. We made an immense fire of pieces of wreck, boiled our kettles and sat round it. It was then pretty dark but the moon rose soon after. Towards morning it came away to blow from the north with snow. We got into the boats and towed the mast off to the ship. We reached her in about three and a half hours. When rowing off I saw the beautiful phosphorescent appearance called the "water burn". On the beach we saw one of the *John*'s iron blubber tanks. On the land were many traces of hares. One of the boats caught many young eider ducks unable to fly....

THURSDAY 1 SEPTEMBER

Fine, clear weather. Four boats inshore. After pulling for two hours we...landed on a flat rocky island. We had coals in each boat with a sort of grate made of hoops. We set our coffee kettles and dined on the rocks. After dinner I and Robert Crombie walked across the island.... The island was covered with long grass, moss, the faces of the rocks clad with the dwarf willow like fruit trees nailed to a garden wall. The trunks of these willows were sometimes so large as to measure one and a half inches in diameter. I found a species of saxifrage and the foxtail grass so common in the bogs of our own country. I caught a small black and red butterfly. I was told by the seamen that they had seen large butterflies and a bee like our yellow earth bee. The island had been much frequented by the Eskimo.... Many of their stone circles, watching stations for hunting, were met with. Their huts, fireplaces, quite blackened with smoke. I found one of their stone lamps. Many seals' and whales' bones were lying about the isle. I observed the skulls of a deer, dog, and a sea horse....

FRIDAY 2 SEPTEMBER

Blowing a gale of wind from the NW, the *Mademoiselle* and *Granville* of Dieppe in company. Towards evening the weather became more moderate. I went on board of the *Granville* for some wine for use [of] our crew.

SATURDAY 3 SEPTEMBER

Fine, clear weather. Four boats inshore, pulled up a fiord for ten or twelve miles. They then landed, lighted a fire. Some of the men climbed up a hill from the summit of which they saw the fiords leading up through the land as far as the eye could reach. They met with many Eskimo graves. In the evening we hoisted our ensign as a signal to the other ships to send letters on board as we thought there was no more to be done....

SEPTEMBER BROUGHT SIGNS of the approach of dreaded winter. Temperatures dropped further below freezing each night, skimming the surfaces of freshwater ponds with thin new ice. Insect voices fell silent among the seer, yellowed, leaves of ground-hugging shrubs and plants. The air vibrated to the wingbeats of migratory birds deserting before the imminent wintry blasts. And in the arctic seas, soon to be mantled by ice, sea mammals retreated for survival, the bowhead whale among them. The sun, now low above the equator, had all but abandoned the arctic regions to darkness and cold.

It was a time of anxiety and foreboding for the whalemen; the consequences of getting frozen in for the winter were deeply feared. Therefore the captain's decision to end the whaling was welcome; it allowed the men to release the fetters from their private thoughts and to anticipate without restraint the familiar comforts of homes, families, and friends. But alas! Captain Allan then changed his mind (because he had come to his senses when the rum ran out, the surgeon supposed), and he decided to persevere a while longer in the pursuit of whales. The vessel remained another ten days on the Baffin coast, but no whales were sighted. Finally, on 14 September off Cape Searle, the boats were hoisted on board for the last time, the whale lines overhauled, and the sea log commenced. The *Hercules* was on her way home, with the blubber and bone of fourteen whales stowed below.

A scrimshaw whaling scene etched upon a piece of whale jawbone shows two boats planting hand-thrown harpoons while the ship in the background dodges among icebergs.
(Hull Museums)

Four

Nothing Seems to Gratify

We are cruising in a country
Filled with ice and snow,
Where nothing seems to gratify
But the whales blow.

Talk at tea was upon strawberries and cream. Our captain has not been at home for
twenty-one years past in their season. JOHN WANLESS, 1834

On 1 May 1834, two and a half years after the return of the *Hercules*, the whaler *Thomas* departed from Dundee, one of about seventy British vessels bound that spring for the Davis Strait fishery. Long periods at sea and endless vistas of ice may have seemed tedious to some of the old arctic hands, deprived of home, family, friends, and the joys of summer, but to the ship's surgeon, John Wanless, the entire experience was exciting and worthy of careful notation in his private journal. A voyage to the whaling grounds the year before had in no way dulled his appreciation of the magnificent scenery, or his curiosity about all manner of arctic phenomena. The near illegibility of much of his handwriting is more than compensated by the thoroughness and lucidity of his description of flensing, making off, and other whaling operations.

As the *Thomas* left the land behind the men were put to work making things shipshape. They unbent the anchor cables and stored them below, repaired parts of the rigging, and overhauled the whaling gear. On inspecting the stores it was found that merchants had supplied sub-standard food for the voyage, a discovery that did nothing to promote optimism and good morale. Suffering at the outset from homesickness and seasickness, the surgeon nevertheless joined eagerly in the first Saturday evening toast to wives and sweethearts "by quaffing a tumbler of rum punch," and soon settled comfortably into the shipboard routine. There was little in the way of official duties, but three weeks out he was given an opportunity to make himself useful when a seaman jammed his hand in a block, requiring the amputation of a finger.

When the ship began encountering drift ice south of Cape Farewell the final preparation were made for whaling. The men hoisted the crow's nest aloft to protect lookouts from wind and cold, and fitted out each whaleboat with harpoons, lances, and two-thirds of a mile of line coiled carefully into tubs. On 24 May they arrived at the Southwest Ice. Wanless learned that "formerly the whaleships went to no other place; the whales being here in abundance they soon filled their vessels in a short time and went home early." But intensive whaling had depleted this region by the 1830s, so the *Thomas*, following a similar course to the *Hercules*, proceeded north along the pack ice boundary until she reached Disko Bay, whose calm, shining surface, dotted with white, sail-like icebergs, provided a

magnificent mirror-setting for the surrounding mountains rising in the distance to caps of brilliant, sun-struck glaciers. No whales were seen in the bay and ice conditions looked favourable to the northward, so Captain Alex Cook pressed on up the coast of Greenland.

On 11 June the ship reached Upernavik, the northernmost Danish settlement on the coast. Wanless recalled that "last year we had a boat's crew of Danes with Eskimo women a day on board. They danced to a fiddler's notes most admirably to the entire satisfaction of the beholders." Pack ice, adverse winds, and uncharted reefs made sailing frustrating and hazardous, and at times there was no choice but to secure the vessel with bent iron "ice anchors" stuck into holes made in the ice and await more favourable conditions, hauling the ship ahead by means of the capstan and windlass when feasible. When not required for

work the men "practised ball...for an hour or so" and hiked ashore to an island "where we saw five or six graves of our own countrymen. They were not covered with turf; nothing but rough stones showed their place of rest, the stones heaped upon one another. They may be quiet enough here till they are called forth at the great day...." At midnight on the twelfth the *Thomas* made fast to a berg in company with the *Eliza Swan*, *Neptune*, and a German whaler, the *Altona*.

FRIDAY 13 JUNE

After the ship was moved to the berg I went with a boat in search of ducks this morning to the Duck Islands. On the way I shot arctic dove; beautiful creatures they are with red feet. Several large flocks rose and fled while the boat

Far left: John Wanless, surgeon of the Thomas *in 1834. (Dundee Museums)*

Left: A sketch in Wanless' journal showing (top) *a boat approaching a whale and* (bottom) *two boats fast with hand harpoons while a third pulls up to assist. (Dundee Museums)*

Right: Upernavik, Greenland, photographed in 1893 by J. W. Allen. (Glasgow University Library)

was not within a mile of them. Bad hopes said I. The boat was made fast to the ice. At our landing I gave a gun to the best gunner and kept my own and we travelled round this island without seeing a fowl. Saw some Eskimo old huts with some old bones of bear's feet scattered about them. The impression of a living bear we saw upon the snow and appeared to be quite recent. I was tired enough before we got to the boat again, and sweating. Some places the snow was so soft that up to the loins we sank, and with great difficulty extricated ourselves; other places nothing but large stones. The shyness of the ducks was attributed to so many ships being before us. Left the island. Another duck shot. Came on board at six o'clock in the morning and went to bed. Did not rise till dinner time, then I heard the sound of another ship's bell at eight bells or twelve o'clock.

It is beautiful weather. The cook and his mate are engaged getting on board melted snow from the top of the berg. It is quite low, large pools are here, and the water is fresh. For the first time I viewed the sun at midnight last and he is a good way out of the horizon.

SATURDAY 14 JUNE
At 6 A.M. the ship got under way with a northeasterly breeze. The ice went off as we beat north. We went ahead of the other ships but had to bring up at 7 P.M. for a neck of ice....

SABBATH 15 JUNE
Last night at twelve o'clock called all hands, stowed the sails, and the men went upon the ice to drag the ship along

the land floe by a rope fastened to the foremast head. Every man is supplied with a rowraddie, a piece of canvas with a small piece of cord at both ends. An eye is through the end of one and a button at the other. The canvas is put round their shoulders while the button is placed on the tow rope and overlapped by its own cord, not to slip, at the same time being through the eye. At two o'clock got through the neck this morning with our three companions and beat north in Melville Bay in water about three miles broad all day, leaving the other vessels. At 2 P.M. saw a ship to windward shut in by a floe to the land ice and they had all hands employed sawing through this floe in order to get out. It is seven and eight feet thick. The ship is the *Lee* of Hull. At four o'clock saw a chest with one end driven out lying upon a small bit of ice. Saw twenty or thirty more ships, some made fast and others working northward. At eight o'clock I discovered from the deck a wreck on the weather bow with the glass, and apparently a number of men upon her. At ten o'clock lowered away a boat into which I went to learn the name and the manner she had met with this deplorable circumstance. When we came to the wreck the *Lee*'s boat was there with the crew plunder-

Left: Ships crushed by ice were stripped of useful articles and burned. When the men broke into the spirits store a "Baffin Fair" usually resulted. (The MIT Museum)

Right: The crew of the Narwhal *sawing a dock in the ice, 1859. Sketch by R. H. Hilliard. (Glenbow Museum)*

ing her. They got very little; so clean had she been pilked [i.e., picked] that not an article we could have taken. The rigging and all the masts (but the fore and mainmast, because they were under water) were cut adrift. She was upon the starboard side. The water was above the larboard side of the main hatchway. She was scuttled here and there below the hold beams. Biscuits seemed to be the only thing left. The water in the inside was covered with them [swimming?]. Last Tuesday, under topmast steering sails between two floes while running north, she was thus squeezed, we were told. The men saved their lives except one who was at the helm. He had been anxious to save his clothes and ran down for this purpose but the poor fellow was smothered in this act by the ship's canting, as we saw her. He was taken out from her broadside with his arms extended and his hands clenching fast their object. The cooper was also taken out opposite the galley, but in existence [still alive]. The captain had not so much time as take his watch. A person who has never seen this country can have only a faint idea of the power these floes possess when under way. A ship appears to offer no resistance to their velocity. Southerly winds, I may mention, have been the cause of the loss of many a good ship in this bay. The ice drives right in upon the land....

MONDAY 16 JUNE

At 7 A.M. called all hands to tow the ship. There was not a breath of wind and the water smooth as a mill pond would be. Large bergs as we passed by re-echoed the many loud songs and wild cries of the men after having rung through their rugged crevices. The other ships ahead have all hands towing and tracking. Their yells were heard six or seven miles. Today we were beat by the companions, our ship being heavy and dull when no wind. The men wrought hard till about six o'clock. They then were relieved by our journey having to be stopped for another neck of ice. The Dutchman [German] was close to us. Some of our crew went on board for gin; the barter was any wearing apparel. Their doctor came on board of our ship and asked for me. I was just going to bed—at 11 P.M. He could speak little English. However, I got the assistance of a Dane we had on board and passed a jovial night with him, so we became remarkably good friends. He was in need of an sharp lancet and some adhesive plaster. I made him a present of two lancets and gave him the one half of my plaster with a piece of linen. I was everything with him and he called me "a fine fellow." He liked our rum better than their gin and brandy. The captain and mate could get no sleep with us talking so loud, I was told. I went on board with him, got [some] of his gin toddy, a Dutch pipe, five stalks of Negro Head tobacco and a quantity of shag. I was pretty far advanced. At the moment all hands were called. This put a stop to our farther friendship and I departed with my booty to my own home smiling, our men all ashore upon the ice tracking. We hoped to see each other again if a time to do so was permitted. The watch showed me it was six o'clock Tuesday morning. In half an hour we got through this neck; the ice slacked and we had a light breeze from the northward. At half-past seven set the watch. I went to bed and was awakened by the noise of the men at the capstan heaving the ship through two large floes at eight o'clock. ...

Rose at 12 A.M.; weather clear and the rays of the sun very hot reflected from the ice, almost like to burn the face. On the east of us there appears to be a whole continent of solid ice eight feet thick. The loose stuff is all straggled about, making it difficult to keep the ship from striking. We continued till six o'clock in the evening working to windward. At this hour there was less wind. All hands knocked out to tow with the six boats. The ships ahead have all hands too. The ship has had some heavy shakes by the ice today. A great number of fowls chirping, called rodges [dovekies]. Men too busy for shooting. At nine o'clock the breeze came stronger from the north-

ward. The boats were called alongside and hoist up. The watch was set at half past nine.

One of the men saw another wreck. I am almost afraid to go to my bed after viewing the state of this vessel. Her masts were torn from the deck and crashed upon the ice. Her deck was torn asunder and every part that would offer resistance was crushed to atoms. The land ice and the floe at their meeting had burst their edges (eight feet as we imagined) with the pressure and forced those clumsy pieces into direful heaps above the general level. One piece was seen suspending part of her beams after it had gone right through her quarter.

All hands called again at half-past ten o'clock to tow. Two of our companions have got away through between two large floes; one is jammed to the land ice. We made an attempt to go through here also, ten minutes after them, but in vain. Some pieces interrupted the passage at first. The point from one was sawn off about thirteen or fourteen yards and pulled astern. At last the whole floes came closer together and there was some of our crew thinking we were in a critical situation (I saw some of our crew running below with their lashings for the chests privately). The process of sawing is conducted by erecting a triangle of three spars about eighteen feet high. At the apex there is a gin or wheel. Over this a rope is passed; one end of it fastened to the saw fourteen feet in length, the other to several ropes. These the men take hold of to elevate the saw, while it is depressed by the harpooners, generally by taking hold of four spokes placed coming off at right angles upon the centre of the top of the saw. One of our boats was stove; she was towing ahead of the ship and got her timbers broke between the piece and the ship's bows.

A S THEY WORKED THE SHIP LABORIOUSLY through the pack ice during the next few days, with more than fifty other whalers within sight, Wanless kept his personal gear close at hand "ready for lifting upon the ice if it blows." By the 20th they had reached a large body of open water but the winds failed them. The men were put to work tarring the rigging while the surgeon, complaining that the lack of small shot on board was "a great pity when among the little birds," blasted the feathers off several dovekies and succeeded in killing eighteen (one to be stuffed and the rest made into pies). Most of the men, he noted, were suffering from eye inflammation caused by the brilliance of the sun reflected from water and ice, and "all our faces are burnt as if we had been crossing the line under the sun. The lips are cracked too." Yet temperatures dropped low enough to decorate men's beards and ship's rigging with filligrees of hoar-frost.

SABBATH 22 JUNE

All hands called at six o'clock this morning to tow. Weather thick. The rigging is entirely white with this freezing, from the deck to the masthead. Watch set at eight o'clock and kept two boats at the tow rope. I saw a burgomaster fly away with a rodge in his mouth. At ten o'clock got in sight of the grand fleet made fast to the land ice. In half an hour after we did the same astern of the *Mary Frances* of Hull. We being the packet [i.e., bearing mail] were soon crowded with men expecting news. I went on board the *Charlotte* of Dundee to see my friend Mr Milne and to deliver two letters. He was six or seven ships ahead of us. Travelled back upon the ice although the boat took me there. This is a scene! About a mile in length covered with ships, every one hauled as close to his neighbour's stern as possible. I was sleepy and went to bed at three o'clock in the afternoon. At half past four my nap was disturbed by hearing all hands called, not to tow or track but to look out for the ship, for two of them ahead of us were in a terrible squeeze by the ice driving to the southward.... *Resolution* and *Undaunted* had all hands sawing a dock. All our helms are unshipped and hung athwart the stern with two tackles. The *Mary Frances* got part of her stern post torn

off by the ice and was in a nip close to us.... Our clothes and boxes are all in readiness. Some ships had theirs upon the ice, in their boats.

MONDAY 23 JUNE

Towards midnight last the ice began to slack and all the ships reached the water after tracking &c. at two o'clock this morning, with the wind from the northward, till nine o'clock. We had sometimes a breeze, at other times becalmed. All hands called at ten o'clock when the boats were lowered and the helm hard a weather, the mizzen brailed up, and other things required to make the wear in a short time. This we were obliged to do by the *Dee* of Aberdeen running right upon us, and a large berg to leeward; we just cleared. The watch was set again half a hour. We kept looking north. Half a mile ahead of us the ships when they reach have to lower away their royals [uppermost sails] for the breeze. We are becalmed. After dinner the mate and I went to shoot, brought on board about 100 rodges in two hours.... We now may sleep with greater ease for Cape York is to the south. This is [the] northern point of that awful bay [Melville Bay].... The ships are all around; sixty-five sail could be numbered. It is beautiful to view so many under canvas with clear, glittering rays from the straggling ice and icebergs shining upon each.

TUESDAY 24 JUNE

Becalmed last night. This morning at seven o'clock we got a breeze from the southward. All the vessels are running with steering sails on both sides, low and aloft. In the forenoon the *North Pole* of Leith was lying to the wind as we passed with his instant at the masthead. This was done to commit to the deep the last remains of a seaman who died in consumption (I heard). The process is gone through with prayers generally before he is launched from a board hung over the side. One end is tripped; the corpse slides down gently and sinks by having hung at the feet sand and shot. The watches have been inverted today—four hours below instead of eight—to get the hold clear in case we should catch some whales soon. There were plenty of fowls today; [but] ship sailing too fast [to shoot]. Cooper knocked down the water casks and made them into shakes to make less bulk. Passed a large glacier, or ice manufactory, today between two high hills.

THE MOUNTAINS OF ELLESMERE ISLAND were sighted at noon on the 25th. In the face of headwinds the *Thomas* and a few other vessels turned southward and arrived off the mouth of Lancaster Sound on the 27th. As the ships approached the most productive area within the entire Davis Strait whale fishery an abundance of narwhals ("a good sign of fish, the sailors say") and fulmars made the lookouts especially attentive. Their vigilance soon paid off.

SATURDAY 28 JUNE

This morning at one o'clock all hands were called and we made what is termed a loose fall, or otherwise [all] six boats were sent to chase whales. A pretty chase some of them got too, before one was caught, which happened at 8 A.M. In two hours she was alongside. After the men took breakfast everything was made ready for flensing. A fall [rope] is rove through two blocks in front of the blubber guy and fastens [i.e., leads] to the fore capstan. Another is rove through two blocks on the after part and fastened [i.e., led] to the main capstan. A block with a hook [i.e., the spek tackles] hangs from each of these falls. Another one stretches to the windlass and is called cant fall. After the fish is made fast by the rump at the fore rigging and at the main rigging by the nose tackle—or a tackle sustaining the tendency of the head to sink by a strap hooked to the block, passed through a hole made with a knife at the uniting of [the] jawbones—the harpooners take their berth, with their large boots spurred beneath, upon the fish while sailors man capstans. Two incisions are made transversely through the blubber covering the lower jawbones. A hole is made with the blubber knife into a corner, a strop [strap] is put through and hooked to the falls block, then there is run round the capstan [i.e., haul in] singing loudly. At the left fin a flap is made; this is fastened to the cant and the windlass is hove round as the fish requires to be turned. Slips [strips of blubber] are taken from the cant piece to the tail. Harpooners are armed with blubber spades and knives. The slips, after they are on the deck, are cut down into pieces about a foot square by the boatsteerers. The line managers then with their picky-haaks throw them down the hatchway through an opening in the hatch that will not allow larger portions admittance. In the hold two men are situated and are clothed with watery poops [water proofs?] and large boots. Each has a picky-haaks with which he stows the blubber properly. These men are denominated king and assistant. They have double drams [of grog] for their reward. The bone [baleen] is taken from the fish as she is canted by the bone gear. This fish's bone measured nine feet in length, that is from the root in the gums to the tip [of the longest slab]. It serves the purpose of retaining the small animalculi in their mouth by being covered internally with long hairs thickly placed while the water is ejected. The blubber was about a foot thick. A black skin upon its outer surface measures generally an inch. The head occupies half the length nearly of its body. The tail is placed horizontally.

The flensing was finished three hours and a half after the commencement of the process, the deck cleaned, and the watch set at 2 P.M. A boat was manned to look out for a chance and be ready; this is called branning. In about half an hour a whale was seen from the deck lulling [lolling, or resting]. The boat [illegible words] after her. The water was very smooth; they had to pull quietly upon her. At last she was struck and they called a fall. We five or six on deck shouted out the same, knocking with capstan bars upon

FLENSING

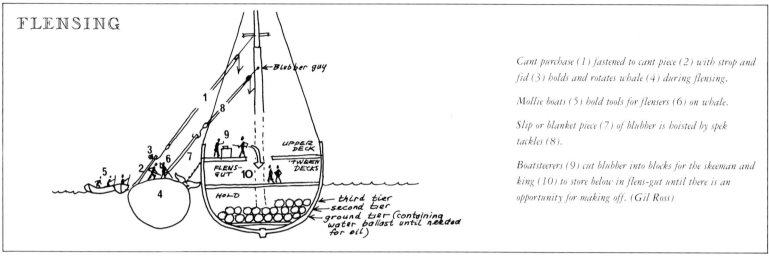

Cant purchase (1) fastened to cant piece (2) with strop and fid (3) holds and rotates whale (4) during flensing.

Mollie boats (5) hold tools for flensers (6) on whale.

Slip or blanket piece (7) of blubber is hoisted by spek tackles (8).

Boatsteerers (9) cut blubber into blocks for the skeeman and king (10) to store below in flens-gut until there is an opportunity for making off. (Gil Ross)

MAKING OFF

Blubber blocks are hoisted to the deck from the flens-gut (1), trimmed by the krengers (2), transported by clashers (3) to the clash hooks, skinned by skinners (4), deposited in bank (5), sliced into small pieces by choppers (6), pushed into spek trough (7), and down lull (8) — controlled by lull-boy with nippers (9) — to the skeeman and king in the hold (10), who store them in casks or tanks. (Gil Ross)

the deck to rouse the men from their newly pleasant sleep. In an instant the deck was crowded with the men running to get into the first boat, with their eyes scarcely open some of them and their clothes in a sling to put on when they got more time. This one stopped about an hour beneath before she came back to blow. They got their harpoons in and, having a good strain, all the boats held on by each other as he ran through the water with them. At last she came to the surface and made a dreadful noise with her fins and tail. None of them suffered any damage farther than having their oars unshipped. She was brought alongside and flensed. The watch set at 8 P.M.

MONDAY 30 JUNE

No whales have been seen through the course of last night. This morning the watch was employed gumming the bone. A plank is placed upon two stools. The men stand before it and scrape with a bent piece of iron the gum

Left: A British whaler under topsails and spanker, with garland aloft, is flensing a whale alongside, attended by mollie boats. The baleen is being hoisted on board. Elsewhere other boats are busy whaling while some men are hunting sea mammals on the pack ice. Sketch by John Gowland. (Dundee Museums)

Lower left: Flensing ("cutting in" to Americans), with the flensers working from a cutting stage rather than the mollie boats used by British whalers. Sketch by Timothy Packard. (Houghton Library)

Right: American arctic whalers extracted the oil from the blubber in tryworks on board ship and stored it in casks. British whalers did not "try out" the oil, but returned to port with the blubber. From the Orray Taft *sketchbook. (Kendall Whaling Museum)*

from one split [plate] at a time. The hair is cut off by another crew and thrown down in the hold. In the afternoon we stood in towards the land. A great many of the ships were made fast to the land ice and to a large floe on its outside. At eight o'clock we also made fast to the latter. The Dutchman inshore and has one fish. Some of the ships are clean. One of our boats was put on the bran. One of the harpooners fell from the plank and got himself severely bruised this forenoon.

TUESDAY 1 JULY

At 4 A.M. all hands were called to make off the blubber in the hold. This is a tedious action. Three men whip the blubber hooked on by the king and assistant upon deck. Two harpooners skin and [remove?] the muscular substance and cellular that had been left on by the flensing people. After it undergoes this it is placed one lump at a time upon the clash hooks before the other harpooners, by a man to each with a [chunk?] in both hands, to be skinned. A large trough put athwart the hatchway served with a lid going upon hinges, and a square hole in its bottom. A frame is adapted to this and hanging from it is the hose. At the middle of the hose two sticks when pressed together can prevent anything from dropping through them. The boatsteerers stand behind this trough

and the portions skinned and cleaned are hove upon their [chopping] blocks by the bankers...to be cut in small pieces not exceeding five inches in diameter, about a foot long. These pieces are shoved down the hose and are received into tubs in the hold. They are lastly put into the bung holes of the casks by eight or ten men under the [illegible word] and skeeman. There is a constant noise kept up on decks and below. If the hold men have too much at one time they sing out to the boy at the stocks to "nip the lull"; hence is he called the lull boy. If they require a fresh supply it is "let loose" or "open the sticks." The boy then cries "scoop off" to the deck folks, or to shove the portions into the hose. The blocks are taken from the tail and fins of the whale for hardness. They lie upon the lid folded back upon stools. The boatsteerers' chopping up knives are in this form: A the blade, B the handle, C a prong for turning over the blubber upon the block.

The clash hooks are

The men clashing have in their hands.

The harpooners' knives:

For carrying and shoveing about the blubber:

Speck trough:

At 8 P.M. the whale was made off. The ship was let loose from the floe, the watch set, two boats on the bran, and the ship standing to the eastward among loose ice.

DURING THE FIRST WEEK OF JULY, whaling, flensing, and making off continued with few interruptions. So busy were the crew that the captain had to send the surgeon aloft as masthead lookout—for eight hours straight on one occasion. The men harpooned, killed, and processed four whales in as many days, all off the mouth of Pond Inlet.

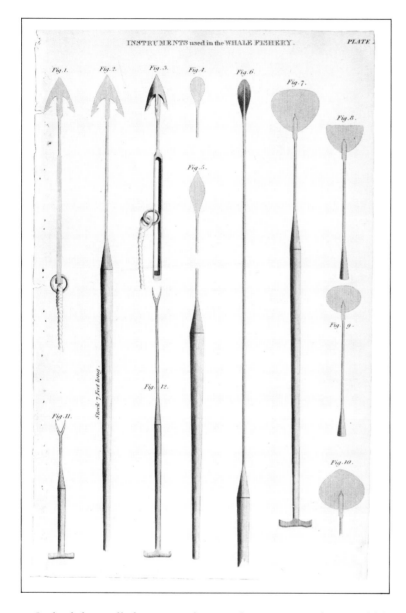

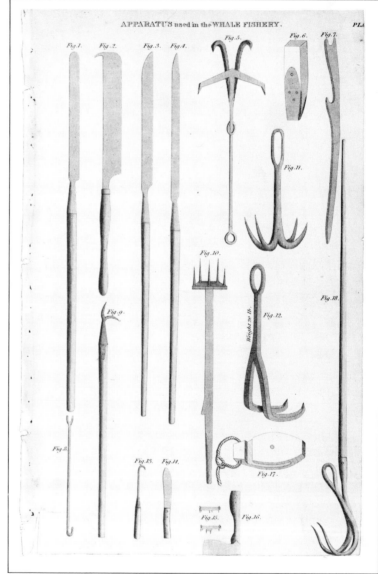

It had been little more than a dozen years since British whalers had begun to sail to Pond Inlet and other localities on the west side of Baffin Bay. They had encountered native people remarkably different from those usually met with at the Greenland settlements. They were Eskimo in race and culture, to be sure, but unlike the Greenlanders they had remained beyond the frontier of European exploitation and colonization for a few more centuries, and were only now beginning to experience repeated contacts with whalers. The following account by Surgeon Wanless is one of the earliest descriptions of the Pond Inlet people.

Engravings from Scoresby's An Account of the Arctic Regions, *showing the various tools used by British whalers in the nineteenth century. (Metro Toronto Library)*

Far left: Fig. 1 – gun harpoon; fig. 2 – harpoon; fig. 3 – gun harpoon; fig. 4, 5, 6, – lances; fig. 7, 8, 9, 10 – blubber spades; fig. 11, 12 – prickers.

Left: Fig. 1 – blubber knife; fig. 2 – chopping knife; fig. 3 – strand knife; fig. 4 – Tail knife; fig. 5 – bone geer; fig. 6 – bone wedge; fig. 7 – mik or rest for harpoon; fig. 8 – third hand (used in flensing); fig. 9 – pick-haak; fig. 10 – clash; fig. 11 – grapnel; fig. 12 – ice grapnel (used in warping); fig. 13 – krenging hook; fig. 14 – krenging knife; fig. 15 – spurs; fig. 16 – axe; fig. 17 – snatch block.

Top right: Fig. 1 – harpoon gun; fig. 2 – boat's wince ("apparatus used in the whale-boats for heaving in the line when a great quantity has been withdrawn"); fig. 3 – hand hooks; fig. 4 – ice drill; fig. 5 – gun harpoon; fig. 6 – seal club ("an instrument by which seals are usually killed"); fig. 7 – king's fork ("an instrument by which pieces of blubber are moved about from one place to another").

Bottom right: An apparatus designed to replace the cumbersome spek trough and produce blubber chunks of a uniform size.

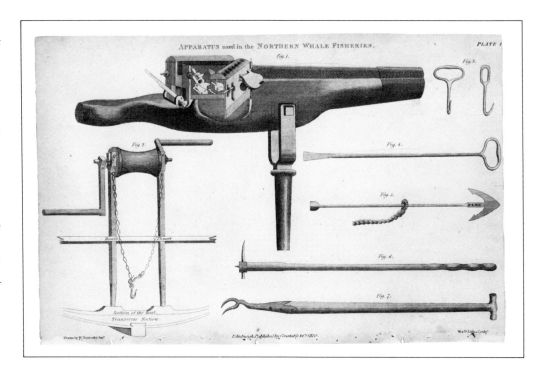

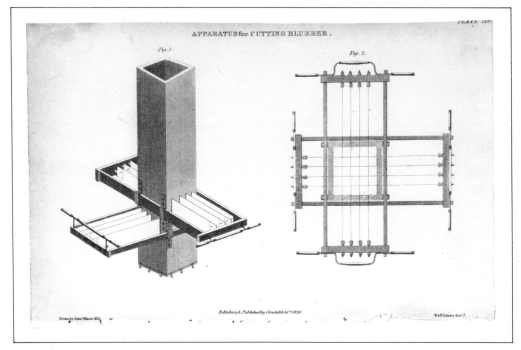

Sketches of Eskimo life from Wanless' journal.
(Dundee Museums)

MONDAY 7 JULY

At 4 A.M. called all hands to make off and flense and two sledges with Eskimo were seen coming to the ship. They came on board and a terrible noise was made endeavouring to imitate the words said by the men. The two women or cunas were dressed in watery poops with a long pendant fore and aft, short trousers, and large boots wide enough to contain a child. The poop is supplied with a hood into which one of them had a young infant. They were tattooed upon the brow, cheeks, chin and thighs, by what I could not learn. They nor the men would taste not spirits. I gave all my needles nearly to them and some thread, as these seemed to be very much thought of. They had not anything to give in return. I fired a gun and all of them ran and put their hands on their ears, very much surprised. The master and I went ashore to view their [mukies?] (dogs) and [slukies?] (sledges). One of the women went along with us. She took from all the dogs pieces of leather that were tied around their feet to prevent the ice from cutting them – this was instead of shoes. The ties were small threads of whalebone. The [slukies?] were made of two planks for sides about twelve or thirteen feet in length, covered at the bottom with bone. Other pieces of wood were fastened across. Upon the top of this they sat. The dogs were all harnessed. Each trace

was fastened to the front by a strip of leather passing through the eye of the trace made of bone; they appeared to be surly tikes and would have bitten us had the cuna not spoken to them. The skin of the blubber was eaten deliciously and the blubber itself was a sweet morsel. They stowed their [slukies?] with what could not be eaten at this time. The whip was so long and the shaft so short that it became remarkable, the striking of a particular dog if required.

They all departed from us – four pikininies, three men and two women – and went along the ice to another ship to the southward of us. They were greatly astonished at looking in a mirror and, observing the contrast, started back holding up their hands. Seals' skins principally afford their clothes but they have deer skins and bears', sewed with tendons. Their cheek bones are high and are far separated, hair long and dark hanging over their shoulders; black eyes and sharp; skin of a tawny colour. Hands and feet are exceedingly small in proportion to their body. The cunas had their heads adorned with ribbons like children's busking dolls the sailors [use?]. They leaped with joy after this was done and brass buttons were put on the foreheads of the pikininies, to their great delight. "Chimo philiday" is their favourite talk.

TUESDAY 8 JULY

All hands called at 8 A.M. to make off. At six four sledges of Yacks came to us, different from the former. They stayed on board two or three hours. I went along the floe about a quarter of a mile with them, giving buttons all the time to keep the peace. I offered a good deal of things for a young dog they had. It seemed to be a great favourite of the cuna belonging to the sledge and for this reason I imagine Yack himself was prevented from making the bargain. Fifteen to twenty dogs in a sledge. One of them had a unicorn's horn made into a lance. I procured it. We had all of them in the cabin eating fat. In the afternoon I got three hands to go shooting. Shot twenty-seven looms.

Weather rather hazy and sometimes raining a little. All the blubber was made by 5 P.M., the bilge water pumped out and watch set at six. No whales seen today. Ship let loose from the ice and stood to the southward with the slight breeze southerly. The Yacks had deers' horns in back of sledges.

ENCOUNTERS SUCH AS THOSE described by Wanless must have occurred at a number of localities along the Baffin Island coast as whaling masters, returning year after year, built upon the navigational knowledge gained by experience and took their vessels cautiously into uncharted fiords in the search for whales, inadvertently meeting up with remote bands of Eskimos who had not previously come into touch with European man. Contact would continue to occur in this sporadic and unstructured manner for some years, until whalemen discovered regions where ships could remain longer, and came to utilise the skills of the Eskimos systematically in whaling and hunting.

After departing from Pond Inlet with the blubber and bone of eight whales stowed below, the *Thomas* coasted southward with other vessels of the fleet, taking two more whales in July with 1,000 pounds of baleen. The surgeon kept busy collecting specimens: two whale's ears; a nine-foot narwhal tusk; three polar bear skins; a bear head; and a number of birds. A bottlenose whale drew his eager attention on the first of August, but circumstances intervened. As Wanless recorded in his journal, "I would have got an excellent shot but the danger of passing the ball through the heads of the crews prevented me from offering violence in this manner to another animal." Commendable restraint!

The whalers moved steadily southward past Coutts Inlet and Cape Henry Kater and on towards Cape Broughton and Cape Dyer. The men of the *Thomas* could often see twenty to thirty ships nearby and there were many opportunities to exchange news and learn about the adventures of the others. A boat from the *Hecla* (the vessel that had taken Edward Parry in search of the Northwest Passage on three occasions between 1819 and 1825) had narrowly escaped destruction when a nearby iceberg began to overturn. Two boats from the *Regalia* had been lost for twenty-four hours. The *Princess Charlotte* had captured a young whale with a grappling hook. The *Grenville Bay*, one of the last ships to leave the Greenland coast that season, had happened upon the survivors of the wrecked *Heroine*, who had been living upon the ice for two weeks. The *Triad* had entered an unknown fiord where (as their crew described it) whales numbered in the thousands but owing to bad luck and rotten lines they had managed to kill only three. At Cape Searle the *Monarch* had met Eskimos who could sever a thread at twenty yards with a thrown lance. The mate of *Middleton* had been killed during a whale chase, and the *Hecla* had buried a man on Cape Hooper. The anecdotes and tall tales, some amusing and others tragic, passed from ship to ship, and the crew of the *Thomas* were able to recount a few misadventures of their own. Ice driving ashore had forced the ship to work offshore to the protection of an iceberg, but the berg had suddenly disintegrated, causing them to cast loose again and pass the night dodging back and forth. On another occasion the captain had broken up a fight among the crew by putting all boats on the bran "to cool their courage." The surgeon, always intent upon securing specimens of wildlife, had once overcharged his gun, which had recoiled against his head and bounced overboard into the sea.

Thus, without any of the radio aids that we today take for granted, the whalemen maintained a sort of grass-roots communication throughout the whaling season by frequent gamming with other vessels. Aside from the natural inclination in remote places to meet and exchange information there were practical advantages to be gained. News about new whaling grounds might result in a more profitable voyage. Awareness of difficulties that had beset other vessels could enable a whaling master to avoid similar circumstances. Knowledge of the whereabouts and destinations of other ships could become vital in emergencies, when co-operative action might be the only hope for salvation.

Viewforth *(middle)*, Jane *(left), and*
Middleton *beset in the autumn of 1835.*
(Hull Museums)

Five

The Grim Tyrant

THE ICE DRIFT of the whaleship *Dundee* in 1826–27 (Chapter 2) and a number of shipwrecks had pointed up the danger of carrying the search for whales into Melville Bay and beyond. Yet, in the years that followed, British whalers continued to press northward along the Greenland coast, each master reluctant to hang back while others attained the rich whaling grounds of the West Side. The shipowners, although desirous that the captains should make every reasonable effort to fill their ships with blubber and bone, were nevertheless reluctant to "waste" money on warm clothing and extra provisions merely for the possibility (a remote one, they liked to believe) that a ship would be beset by ice at the end of the whaling season and forced to spend the winter in the drifting pack. If a vessel had the misfortune to be trapped in the ice her crew would simply have to stretch their food resources a little more thinly and supplement the shipboard diet with game. After all, had not the men of the *Dundee* – a ship outfitted in the usual manner for one summer only – successfully endured the arctic winter in 1826–27? Had they not emerged from Davis Strait in spring to rejoin their families and take up whaling again

in subsequent seasons? Surely their commander had been right when he declared that if there was no let up in the resolve of the men, a crew could safely endure a winter in the ice. The owners continued to place their faith in that excellent Victorian commodity – resolve. It was ever so much cheaper than provisions and clothing.

Subsequent ice drifts would demonstrate what the shipowners did not then appreciate, that the crew of the *Dundee* had been, in one sense, lucky. The ship had drifted down the eastern part of Baffin Bay, where the climate tends to be milder than on the West Side. As the ice carried them south towards the margin of the pack small pools and lanes of water had opened up among the floes, and bowhead whales had appeared in them. The men had managed to kill two, and had benefited also from bears, foxes, and sharks attracted to the whale carcasses. This supply of fresh meat, obtained at a time when each man was reduced to an allowance of only two and a half pounds of bread per week with a little salt meat, had held at bay the whaleman's worst enemy, scurvy.

But what would happen if a ship were beset for the winter in the middle of Baffin Bay or against the Baffin Island coast?

Would whales or other sea mammals rise within the pack? Could bears and foxes be secured for the galley pot? No one knew the answer, but in any case there were many who chose to believe that, even if conditions did turn out to be less favourable for human survival, those heroic qualities of "intrepidity, resolution and perseverance" to which the *Dundee*'s captain had given so much credit would carry the men through safely. Whalemen were given many opportunities to test this supposition in forthcoming years and unfortunately it proved false.

Almost seventy ships sailed to the Davis Strait whale fishery from Scottish and English ports in the spring of 1835, their crews eager to get at the business of slaughtering whales and hopeful of securing full ships and earning generous amounts of oil money. But the arctic ice was at its worst and the season was to become one of the most disastrous ever for the British whaling industry.

In mid-May several whaleships were held up by dense pack off Rifkol—unusually far south—and after reaching Disko on 20 May they experienced great difficulty in advancing farther. They persevered for a month and a half, during which two ships were crushed, and on 12 July, a time when whalers were usually "among the fish" off Pond Inlet and Lancaster Sound, the whaling fleet was still on the Greenland coast near Upernavik, unable to make any headway across Melville Bay. The captains decided to try elsewhere and took their ships 400 miles south again, where some succeeded in crossing Davis Strait to Cape Dyer, Baffin Island, on 27 July but again all efforts to get north to Pond Inlet were thwarted by pack ice, which destroyed another vessel in August.

In early October, well past the usual date of departure from the whaling grounds, a dozen whaleships remained in the tenacious grip of the pack between Home Bay and Cape Dyer. As winter's cold descended, freezing up open leads and holes with new ice, approximately 600 men drifted slowly southward, doing what they could to saw, tow, track, and warp their floating prisons eastward toward open water and freedom. The Hull whaler *Alfred* broke out during the month of October and hurried home with news of the predicament, but by the time she

reached port the Davis Strait pack had claimed two more ship victims, one of which (the *William Torr* of Hull) lost all hands to cold and starvation in the gruesome days that followed. By late November nine ships were still in the ice, beyond the reach of help.

Evidently the whaling interests had learned nothing from the experience of the *Dundee* a decade earlier. As the Hull owners admitted in a pathetic plea for Admiralty assistance in December, the whalers were "not sufficiently provided with the means of supporting life through the severities of an arctic winter." If provisions were insufficient for the ships' normal complements how dreadfully inadequate they must have been for the crews as they then were, burdened with more than a hundred survivors from several wrecks. With financial support from whaling and commercial interests, the Admiralty responded swiftly, dispatching Captain James Clark Ross in charge of HMS *Cove* towards Davis Strait in January 1836, but the most he could hope to accomplish was to provide assistance to any ships that might emerge from the ice, if he could manage to locate them.

The *Viewforth*, a 289-ton bark from Kirkcaldy commanded by Captain Oliphant, was one of the ships enclosed by pack ice on the Baffin Island coast in late July. Through August, whenever open water appeared among the drifting floes, Captain Oliphant tried to work the vessel northward, but his efforts achieved little. Twice, as he attempted to manoeuvre the *Viewforth* closer inshore, the ship grounded on reefs. On the second of these occasions the pack ice drove them more firmly upon the reef and put the ship in grave danger. The men brought up their bedding and sea chests, and with great difficulty got them to the land, expecting the ship to be destroyed momentarily. Happily, she was not destroyed. But in getting her off the reef with the tide the crew lost a windlass, two boats, and several lines.

William Elder, author of the narrative below, was one of the *Viewforth*'s officers, probably the mate. He came of an enterprising Scottish family already involved in publishing (Smith, Elder and Company) and soon to extend its energies to mining and trading in Australia. After his arctic experiences William appears to have left whaling. In 1840, by which time he had become captain of an emigrant ship, he joined his brother, Sir Thomas Elder, and other brothers in the antipodes.

Immediately after the *Viewforth*'s return Elder's shipboard journal was edited by Reverend J. Bain of Kirkcaldy and published in Edinburgh as a small book selling for sixpence. It is perhaps a tribute to Elder's writing style that Bain presented many of his passages *verbatim*, although he omitted a number of daily entries and tended to select those that emphasized religious faith. The account below is taken from Elder's original manuscript journal.

"It is astonishing to see so much ice in the country," observed Elder in late July. A month later he observed, "We are now completely hemmed in with the ice as we can neither get to the northward or can we get out again to the southward." They were in Brodie Bay, along with the *Jane* of Hull and the *Middleton* of Aberdeen.

---

3 SEPTEMBER

This day begins with fine weather. At 2 P.M. I harpooned my first fish. I got a fine stroke at her but after taking about eleven lines they broke. The ice was in a very bad state, so much so that the men could not get over to kill her. She was a fine fish, I suppose about fifty butts. This has been my first performance but I hope I will have better luck next.

15 SEPTEMBER

These twelve days have been extremely cold, the ship lying at the land floe completely frozen up, the bay ice about thirteen inches thick around her and no water to be seen from the masthead. We have harpooned another fish and after taking about nine lines the harpoon drew. We have seen a great many fish, sometimes fifteen at a sight.

16 SEPTEMBER

This day commenced with fine clear weather, the frost very severe. About twelve midnight the fish were that numerous round the ship that the sea was perfectly in motion with them. They were raising the bay ice about eight inches thick under our quarter. We harpooned one. The lines broke after taking out seven lines. That is three heavy fish we have lost now – very, very unlucky. At about 1 A.M. there set in a heavy swell from about ESE and broke up the land floe. We immediately got beset amongst the loose ice. However we got hold of a berg next night which was aground in about thirty fathoms.

1 OCTOBER

These fourteen days we have been in chase every day after day. The mate harpooned another fish but lost her, which now makes four, all averaging as nearly as possible between fifteen or twenty tons each, and had we only got these four we would have had a very safe voyage. We have never heard any news from the rest of the ships, which is very remarkable. It was never known to have happened before.

It is now October and what can we do? The water is nowhere to be seen from the masthead, nothing but a body of ice. The three ships of us – the *Jane*, *Middleton* and us – left Cape Broughton on the first to seek our road out to the eastward. The first day we got a good way out by milldolling – a sort of barricade rigged at the ship's bow for breaking the ice; it consists of any heavy thing so as you can haul it up and down tackling. We got out as far as a berg that we had seen on the third. It was a terrible mass of ice. It was higher than our ship's masthead and it was aground in forty-five fathoms water, about 270 feet below water and about 100 feet above the water. We lay at this berg for six days. We saw a ship to the offing running down the cant [edge] of the ice and we could even see the water. We expected to get clear with very little trouble but how vain are the hopes of man. We cast off from the berg on the ninth and got the ship beset amongst bay ice. It came on a heavy snow during the night and when morning came in there was nothing but a solid sheet of ice

Left: Elder's journal contains this sketch showing the Viewforth *off Cape Micklesham, Baffin Island, in mid-December 1835. A dock has been cut in the floe and whaleboats are on the ice. The* Jane *is visible in the distance. (Scott Polar Research Institute)*

Right: The bleakly beautiful coast of Baffin Island. (Parks Canada)

as far as the eye could reach. We drove down there with the ice to the southward and on the eleventh there came a tremendous gale of wind from the northeastward with drift and snow. We had made the three ships fast to heavy piece of ice. In the hardest of the gale we were driving south about two miles an hour right down upon a ridge of bergs that you thought it was not possible for the ships to go clear. At this critical junction the *Jane* of Hull was lifted two feet out of the water by a pressure of ice. The *Middleton* had a pressure upon her, the ice squeezing up as high as her channels. Our ship got some severe contusions but a gracious Providence ordered it otherwise. We drove through betwixt two bergs that a line would have almost reached from the one to the other. Whenever we got clear of the bergs we found the ice more slack or open but the gale still continuing we drove right down at the entrance of Merchants Bay. When it became calm the heavy frost set in and completely froze the ship up. We are now lying in a condition that there remains very little prospect of us getting out indeed. My mind is made up for a winter in the

arctic regions. The worst of it is we are very short of provisions and so are the other two ships. We are now on a biscuit and a half a day and half-pound beef and about a half-teacupful of meal. The cold too is very intense. The ice at the top of my bed is about a quarter of an inch thick. The decks below are all covered with ice. Indeed the frost is that severe that you can't walk the deck above half an hour at a time. We have prepared the twixt-decks and made it pretty comfortable with sails &c.

15 OCTOBER

We have this day seen a ship from the masthead lying beset amongst the loose ice. She bore about E by S. There were five men left the *Middleton* and three of our men to walk over the ice to her but the distance was too great. They could not reach and – dark night coming on – they returned, but severely frostbitten the most part of them. There was one of our men had fallen in. The frost was that strong that he had not the power to move his limbs. He had to be carried on board. There was another poor fellow

belonging to the *Middleton*. When they had taken his boots off part of the flesh came with them. We have seen these two or three nights a beautiful comet star. It bears about NNW of us. It has a very bright tail.

THIS AND SUBSEQUENT SIGHTINGS of the brilliant comet, whose reappearance at intervals of three-quarters of a century had been predicted in the seventeenth century by the English astronomer Edward Halley (and which will be visible again between December 1985 and April 1986), must have greatly intensified the mystery of the desolate scene around the three ice-bound ships.

By mid-October the dimensions of the crisis were clear to the crews of the *Viewforth*, *Middleton*, and *Jane*, beset together between Cape Broughton and Cape Searle. It was now certain that several months of detainment lay ahead through the coldest part of the year, and that they would have to withstand severe cold with inadequate clothing and on greatly reduced intakes of food. Ill-equipped to venture outside, they would have to endure long, dark, cold hours of discomfort, inactivity, and boredom inside the cramped, frigid, wooden hulls. Yet vigilance,

strength, and endurance would be necessary in order to cope with icebergs and pack ice, which could crush the vessels or drive them upon the reefs and headlands of Baffin Island. And the crews would need to retain enough initiative and energy to hunt, because health, and indeed survival itself, might depend on their success in securing the meat of wild animals.

On board the *Viewforth* the men did what they could to prepare for the approaching winter. An iron stove was built to provide heat below. It certainly made things more comfortable, but not to the point of luxury. Elder related how he would awake with his pillow frozen firmly to the head of his bunk "owing to our breath heating the ice when we are in bed and melting it." "Whenever we rise," he wrote, "the frost immediately takes hold of the damp. There is nothing but ice whichever way you cast your eye in the bed." When snow fell in late October it was wisely left on the deck to "help keep the frost out." The men set traps for a fox seen on the ice, but unfortunately had no success. To alleviate boredom and stimulate morale a school was started for seamen; church services were held every Sunday; and Elder, a deeply religious man, took it upon himself to go among the men each day and read from the Scriptures for a few hours. All these activities were well attended.

The strict rationing of their meagre stocks of food caused discontent among the crews, paticularly that of the *Middleton*. On 27 October her master thought of fleeing the vessel because his crew had "sworn mutiny against him." This news was of grave concern to the officers of all the ships, and Elder gloomily predicted that when the *Middleton*'s food ran out her men would attempt to take by force some of the supplies of the *Viewforth*, and "it will end in nothing but man to man for his dear life." Circumstances, however, were soon to force all crews into a mutual sharing of provisions.

As the ships moved slowly southward with the inexorable drift of the pack the mountainous coast of Baffin Island must have seemed tantalizingly close. Could it not be reached over the ice? From its high summits one might be able to see lanes leading to open water. In its valleys there might be caribou to shoot, or Eskimos from whom meat could be obtained. Its beaches could supply driftwood for fuel. Inspired by such possibilities, five men left the *Jane* on 7 November and set off for the land. But high mountains are entirely capable of deceiving human eyes as to their distance, and apparently unbroken expanses of ice viewed from the deck of a ship may turn out to contain a multiplicity of pressure ridges, open leads, and snow-covered crevasses. What seemed to the sailors of the *Jane* a feasible day's outing proved to be exhausting, hazardous, and, in fact, quite impossible of attainment. They struggled on as best they could but as the short daylight period drew to a close the cliffs seemed as far away as ever. The men turned back and with great difficulty succeeded in reaching the ships by eight in the evening, one of them severely frostbitten in the feet. After this it seemed futile and foolhardy to make for the land.

The southerly drift continued. By 12 November the three vessels were off Durban Island. They had already been fast in the ice for two months.

SATURDAY 14 NOVEMBER

This has been an extremely mild day, clear and foggy throughout with very little wind. We have driven to the southward these twenty-four hours about four miles. We are getting close down on the range of bergs now. We are still hanging on to the floe yet. The *Jane* of Hull has come very close to us this afternoon. There were six men came over the ice today to carry over the poor fellow that got his feet frostbitten. He is still very bad yet, but there are some hopes of his recovery. They told us that their ship has got a severe smashing. They had their provisions and clothes on the piece of the ice expecting every moment their ship would be lost. They got plank and nailed it on outside on the place where she was stove, and find her now to make very little water. The *Middleton* had her sternpost almost squeezed out of her too. It has come on a strong breeze from the NNE tonight which is drifting us fast south. At 10 P.M. it has come on a tremendous gale with a heavy fall

of snow. Our ship has been lying safe however all night.

SUNDAY 15 NOVEMBER

This day has been an another awful eventful day. The wind did not take off till about 9 A.M. when it moderated a little. What a scene was presented to our view when daylight came, in the wreck of the *Middleton*. Oh, I cannot express the feeling that went to our hearts when she was seen. Every one regarded one another in mute despair not knowing and in all likelihood to be our turn next. There were six of our men went over to assist them as they saw a boat coming towards us. What a melancholy tale they brought back. They were all intoxicated and one of them was drowned almost within reach of them. He fell into a hole, the brash ice met over him and he disappeared in a moment forever, to give an account at the ressurection of the deeds done in the body. What an awful warning is it to us when we think of a man going into the presence of his Maker, hurled out of this world in such a terrible state, and on the Lord's most holy day too. O Lord, what can we expect of thy hand? If thou had punished us according to our deserts long before now, thou wouldst have cut us off. And O how thankful ought we to be to thee, O omnipotent God, for having sustained us through this awful night and still spared us to be living monuments of thy mercy. Throughout the rest of the day we have had a strong wind and sleet. There are about thirteen men come over from the wreck altogether; there are two aboard of us. We are still lying at the floe yet with the *Jane* of Hull. We have driven south a long way. We have passed the range of bergs....

MONDAY 16 NOVEMBER

This day has been more mild and less wind. We are still lying at the floe...about four miles off the land. There have never been any more people come over from the *Middleton*, today yet. The report from the *Jane* today is that there is another man drowned. Indeed we suppose the one half will be dead with the cold. The wreck is about three miles off us....

TUESDAY 17 NOVEMBER

The air has been colder today than it has been for this day or two. We had fresh breezes from the NE for the first part but towards night it came on a perfect hurricane. The *Middleton*'s people have got all on board the *Jane* today, some of them severely frost bit. They have saved hardly an article. They are fifty men of them altogether. We are to get twenty-five, the half of them. We have been driving very fast to the southward this day.... We suppose ourselves to be off Cape Dyer. We are still lying at the floe yet with one warp ahead. The *Jane* of Hull is still lying at the same floe about a quarter of a mile from us. The wreck of the *Middleton* is still above water a quarter of a mile to the ENE from us.

WEDNESDAY 18 NOVEMBER

It has still blown extremely heavy all this day likewise. It comes in at intervals – awful gusts fit to shake the mast out of her. We have been driving to the southward along an iron-bound coast all day about three and a half miles an hour. At 10 P.M. the ice drove us in with the land. It was a terrible night, the great towering mountains frowning above you as you saw them dimly through the darkness of the night and expecting every moment that your ship would be dashed upon them. The same time too the ship sustained a very heavy pressure (we were afraid she was gone). We got our provisions and clothes ashore on the ice immediately. It was an awful night. But thanks be to the Lord for his wonderful mercies, for if he had not been with us and sustained us through so many dangers, long ere now we would have been lost. We supposed the land to be Cape Walsingham. We have lain quiet the rest of the night, still driving fast south.

THURSDAY 19 NOVEMBER

We have been still lying quiet all this day although we were driving very fast to the southward. When daylight broke this morning no person would have believed the road we came through the last night. There were two very large icebergs we came inside of. They were that close to the shore that you thought it was hardly possible for us to come through them. It has still blown a severe gale all day....

FRIDAY 20 NOVEMBER

We have had better weather this day. We are still driving fast to the southward yet. The wind has been inclining mostly off the land. The *Jane* of Hull has drifted a long way from us this last night. There were about fifty or sixty men went over from her today to the wreck. I suppose they would get a good deal of her provision as the weather was so fine. We saw a whole tribe of Eskimo travelling to the southward on the land ice in their sledges. They were a long way from us, as much as we could see them from the masthead. It came on dark so soon, which hindered us from seeing how they came on. The land that we were driving past now is awfully romantic, which consists of nothing but tremendous mountains which present to the eye a thousand fantastic forms such as the appearance of ruined castles &c. I rather think the land is only Cape Dyer yet. The day now is very short and the moon away from us likewise, which makes our situation very gloomy; eighteen hours darkness makes a weary night. I had a remarkable dream this afternoon, I thought my arm was sorely cut and it was bleeding profusely. I thought my mother was bandaging it when Mrs. Millar passed by me and she cried out awfully that her son John was dead, I thought I then saw Mr. Millar looking out of a window, a most awful diseased-like countenance such as despair would [induce?]. There have been a great many raven flying about the ship all day but they have never been within shot. The cold has not been so severe today although it is still very severe. I wish very much we had the short day over that so the sun would be meeting us again.

SATURDAY 21 NOVEMBER

...We have never seen the sun for these five days as the sky has been very cloudy, which hinders us from knowing our latitude. We have still two of the *Middleton*'s people aboard of us yet, the rest of them waiting on board the *Jane* to see if they can get any more provision from the wreck.

SUNDAY 22 NOVEMBER

Another Sabbath has come round to us, which the Lord has been pleased to add to this frail piece of mortality. How thankful we ought to be for his unbounded mercies in preserving us through this last wreck; the dangers that we have come through no man would credit. We had prayers aboard of our ship today. It was unanimously proposed by the men that we should all meet in the half-deck and give thanks to that God who had so mercifully delivered us through the past week. I officiated after praying to that God who alone searches the hearts that he would give me a portion of divine grace in taking upon me so awful and so an important matter. Our text was "Bless the Lord O my soul and forget not all his benefits." I took it out of the *Retrospects*, a good book. The sermons in it are really adapted to our situation. We sang the twenty-ninth Paraphrase and part of the 107th psalm. It was truly sublime to hear our voices ascending to the throne of the Most High in such an awful situation and I am sure a great many sang from the heart.

Our ship has lain very quiet all day, the loose ice being congealed into a floe around her. We have driven south very little this day. We got the sun today. Our altitude was 3°30′ which made our lat. 66°16′N. The sun is indeed very low now. I am afraid that we will soon lose her altogether. Scurvy has broken out amongst us too. There are no less than five cases and two of them very bad, their legs literally black. The *Jane* is much the same distance from us today as yesterday. We have no communication with her in these days. The weather has been extremely fine but the air has been very cold all day.

MONDAY 23 NOVEMBER

...I went over the ice about two miles today to meet and give assistance to the *Middleton*'s people. They were twenty of them (a hard blow for us). They have not one bite of provision with them, which will make us very short. They say they have some bread aboard the *Jane* which we intend to go over the ice tomorrow for. There is nothing but starvation staring us in the face now. We are only on a quarter-pound beef for twenty-four hours and the half of a quarter-pound pork and three pounds bread. We are a long way to the southward now. We are in the open of Exeter Bay....

TUESDAY 24 NOVEMBER

This has been likewise a very fine day but the frost still very severe. There were about thirty of us went over to the *Jane* today to assist in getting over the provisions belonging to the men that came aboard yesterday (the distance was about four miles). We started at about 9 A.M....and it took us till one o'clock to get there. We left the *Jane* again about half past one with fifteen bags bread, which made one bag to two of us to carry turn about. We all reached the ship about 5 P.M. but four men that took a different road over the ice, which we have never seen nor heard anything of yet. It was an awful undertaking. I was that much fatigued that I was going to give it up and lie down but the fear of night coming on fairly sent one leg past the other. The worst of it was it was all loose ice and the wind being off the land loosened it which made our way twice as long. We got three hundredweight seventy-three pounds bread out of the *Jane* that had been saved from the wreck. We are still driving past to the south yet. The *Jane* is about a quarter of a mile nearer us today, being drifted closer to us by the tides. It is now twelve midnight and the four poor fellows that were with us have never appeared yet. I am much afraid they will be frosted if not worse. The night is so terrible cold they must be on the ice all night as the last time we saw them they were too far from the *Jane* to get to her.

WEDNESDAY 25 NOVEMBER

...When daylight came in the four poor mortals were seen on the ice. We immediately sent hands over to assist them. There were two carried on board in an awful state. Their boots and stockings had to be cut all to pieces before they came off and their feet were actually frozen to the degree that you could not move one toe. They were that hard that when they walked forward to the galley their bare feet resounded as if they had on wooden shoes.

[The ship's surgeon later reported that one of the men, "when carried to the fire...was not satisfied with being near it, but...actually thrust his feet into the midst of it...."]

We have been employed all day in cutting out a broadside dock into the floe that we are still lying at, but it is a laborious job owing to us parbuckling the most of it upon the ice.

THURSDAY 26 NOVEMBER

This has been another remarkable fine day but a very severe frost, the sky very clear. We have had a very busy day of it in still cutting our dock but the frost being so intense we are making little or no progress, and the men are all complaining too of pains and colds &c. that I think we will make nothing of it. We have got a good large hole cut out alongside by sinking some pieces [of ice] and parbuckling others on the top. The moon was seen tonight again, a welcome messenger to help to cheer us in our darkened paths. The aurora borealis were really awfully grand tonight and very brilliant. The whole sky was illuminated with all the colours of the rainbow. Burns' verse came into my memory, which is very striking: "Or like the borealis race that flits ere you can point its place."

The two men are terrible bad with their feet—the most shocking sight I ever saw. The skin has come off one of their feet and there is nothing but raw flesh. I am much afraid they will lose them. The rest of our men that are ill with the scurvy are still much the same. The *Jane* is a good way from us now to the SE, about five miles or so. My bed is in a terrible state with ice about two to three inches thick on the top. It is a great wonder I am able to stand it but I am in the hand of a Higher Power who is able to keep me up. Our provision is another thing too, being so short I can say that a man needs to have a great command over himself to leave off eating when you are as hungry as ever. The thing is we have a week's bread served out to us at a time, only three pounds, about ten biscuits—about a biscuit and a quarter a day. It needs a good deal of decision of character to sustain from eating any more when you have it in your power.

FRIDAY 27 NOVEMBER

We have had a most intense frost this day. The cold is really past bearing. The air has been very clear and hardly a breath of wind. We are driving slowly to the southward now. We were not able to finish our dock today. The cold was too severe for us being able to work any. The holes that we cut yesterday are all solid again, the ice being about nine or ten inches thick. The *Jane* has driven a long way from us this last night. She is now about six miles from us. Our invalids are still no better and a great many more complaining.

Six P.M., this is a most beautiful night. The moon is risen. There is not a cloud to be seen in the [clear?] blue sky. The last rays of the setting sun are now disappearing (the ice reflecting the light still) for the sun went down four hours ago. Not a hush nor a sound to be heard (for there is no wind). The terrible mountains covered with perpetual snows, the tremendous precipices, the mountains of ice around in all the different shapes you can imagine. The rising moon, the clear blue sky, the stillness

(Barry Ranford)

of the evening, form a contrast awfully sublime. The scenery is really grand and magnificent. It is just such a night for one to contemplate and wonder on the mighty works of God who created all things, and to lead our imagination to things above. And sometimes too my thoughts will steal towards home and my once happy fireside, and the tears too will sometime come rolling down my cheeks, when the thought comes into my head that perhaps I will never see that happy fireside again and all my endearing friends, when I think too on the changes that will and must have taken place since I left you all and how inconsolable you will be for me not appearing at home at the usual time. Sad, sad thoughts – my pen is past describing them.

We have got another two shipmates in the cabin along with us now, the doctor and mate of the *Middleton*. They appear to be very sociable and friendly people. There are seventy-two souls on board our ship now, a great many men. The aurora is very bright just now. It is really as beautiful as I ever saw it. There was likewise a fox seen today close to the ship's stern but it soon disappeared.

SATURDAY 28 NOVEMBER

This has been another remarkably quiet day, hardly an air of wind but the frost awful severe. We have driven off south this last night. The *Jane* is still far away from us today. I am afraid we will be parted altogether, she being amongst more broken-up ice which is spreading faster than this heavy floe that we are frozen up in. The two men that got so sorely frostbitten are no better. Their feet are swelled to an awful degree. I really can't look at them. They are two Orkney men. There is one of our men really very bad with the scurvy. He is past walking any. The other four are on the mending state. This is just such a night as it was last, a glorious evening, but the cold rather more severe. I have had a busy day of it in weighing out coffee and sugar to the men. We have got one pound coffee, a half-pound tea, and two pounds sugar to each man. We have about fifty pounds coffee yet, which is a great benefit to us, (but the tea is all done) and about fifty pounds sugar. We suppose ourselves off Exeter Sound, about fifteen miles to the northward of Cape Walsingham. I wish we were only down past the Cape.... I have some faint hopes that if we keep our ship we will drive down to the sw and get clear there, perhaps in the spring of the year. But that is only a conjecture.

STEADILY THE COLD became more and more intense. William Elder, wearing two shirts, two pairs of socks, and two pair of breeches, was never warm. At night his breath frosted the top of the blankets stiff, and ice lay thick around his bunk. As constant reminders of the perils of venturing beyond the ship the frostbite victims sank hopelessly towards death, which mercifully claimed one of them on 3 December. The crew cut a hole in the ice and dropped the canvas-bound body into the cold, black water. Then the men assembled for prayers, a psalm, and a reading from the Bible.

In the midst of danger and in the shadow of death Elder turned eagerly towards an appreciation of beauty in the forbidding scene around the ship:

It is not in mortal man to conceive such a beautiful scene as that around us. The very land seems sunk in repose and appears to rest more heavily on its foundation. Let a person fancy himself in an immense plain, to stand in the centre and look around him as far as the eye can penetrate and the whole to be filled with little hillocks of ice whiter than marble, all in the most grotesque shapes, even then he will find it easier to conceive than express the grandeur of such a scene.

There was no improvement in the situation of the seventy-one men crowded on board the *Viewforth*. Hunting and trapping had secured the pitiful bag of one fox and two birds, and gradually the inadequate diet of aged shipboard provisions was draining away the strength of the men, and inducing the ominous spread of scurvy.

TUESDAY 8 DECEMBER

We have had cloudy weather and squally, the wind at SE by S, the ship still driving fast to the SW along the land. We are now driving at 3 P.M. past a stupendous iceberg. Arthur's Seat [in Edinburgh] arranged in one of its most wintry garbs is hardly to be compared with it. We are about a quarter-mile from its inside. We have seen the *Jane* today again, which helps to comfort us a little. We have not got the sun yet. I suppose it will be only a degree high now.

Six P.M., it has been the Lord's will to call another of our shipmates to the world of spirits. O, how many, many warnings we are daily getting of holding ourselves in readiness to depart and to be with Christ. What a time has been allowed us for repentance, time in which we will need to give an account of at the great day of judgement.... He was only twenty-two years of age, just in the prime of youth. He had caught a cold sometime in March which brought on consumption and he slipped away. Tonight at six o'clock one sigh was all the struggle the King of Terrors forced from him. I read these two or three days past to him, which he appeared to be very fond of, but I could get no word from him how he felt within. We all assembled in the half-deck, after sowing him up in one of his blankets (the sailors' winding-sheet), for prayers and praise. We sang the third Paraphase and last four verses of the sixty-second psalm. After that I read a sermon out of Dr. Watt. The text was in Job, fourteen chapter, verses 13, 14, 15: "O that thou wouldst hide me in the grave" — it was really a very impressive sermon. The head of the discourse was that the grave was God's hiding place for his people. I thought it was a suitable time for prayer when all our minds were so deeply impressed in seeing the death of one of our shipmates, for a word in season there has a great deal of weight with it, which I have no doubt but a great many present felt so.

This has turned out a most beautiful evening, not a cloud in the sky nor a breath of wind. Every thing seems to be at rest. The moon is on the wane now too. It is really a night for contemplation. Eternity is in these moments.

WEDNESDAY 9 DECEMBER

The first part of this day light airs and cloudy, the ice making a tremendous noise around us, roaring like the loudest thunder. You will see it turning up near twenty feet, crumbling and squeezing into atoms. It is really dreadful and enough to make the stoutest tremble. We are still driving fast to the southwestward yet. At 10 P.M. we committed the lifeless body of our shipmate to the merciless deep, till the sound of the trumpets, when our blessed Saviour shall come in all his glory and rouse him from his slumber with "Restore thy dead thou sea." I read the service. We all assembled around the bier. After launching the body into the deep we all met in the half-deck and had sermon on the shortness of human life, after singing the seventh Paraphase. It was indeed a scene that wrought greatly upon the imagination and the tears were standing in a great many eyes. It came on tonight a heavy gale of wind from SSE with snowdrift. The ship appears to be driving fast down along the land.

THURSDAY 10 DECEMBER

This morning at four o'clock it blew very hard with a blinding sleet, the ice roaring and crashing in an awful manner. It would be indeed a difficult thing for the human imagination to conceive the sensations produced upon our minds by our solitary ship drifting and working her way through regions of eternal frosts. Whole fields of ice as far as the eye can reach, along a coast which presents nothing but desolation in one of its most awful forms, frowning cliffs behind which are glaciers as far as the eye can

penetrate – no home there for us if we were happening to lose our ship. Within fifty paces all around us the ice has squeezed as high as our ship and still the goodness of providence has delivered us. I may say with all safety that certain destruction was within fifty paces of us. It fell calm about six o'clock and has continued so all the rest of the day....We have been employed this day in airing and cleaning the ship, which makes her a great deal more comfortable below.

I have just now left the twixt-decks where I saw a most shocking sight. The poor lad that got his feet frozen is in a bad state. His feet are falling away by the ankles. They are rotting off; the smell they have is dreadful. Gangrene has commenced its direful operations, and the surgeons are afraid of his life. Scurvy has not spread any more amongst us yet, which is very pleasing for us. We have only the three cases yet and little appearance of them getting any better. We had service tonight again in the half-deck. It is really pleasant to see how eager the men are for it. The text was "O that men would praise the Lord for his goodness and his wonderful works to the children of men."

FRIDAY 11 DECEMBER

This has been a very fine day but the frost very strong. We have driven south a long way this last night. We see Millers Island now. We are only a small way to the north of Cape Micklesham, a remarkable bluff headland. The tops of the mountain reach to an awful height....Our invalids are still no better today. I was in the crow's nest today and saw the *Jane* a long way from us. She bears about W by N. The ice appeared to be opening in a good many places as there has been a light air off the land all day.

SATURDAY 12 DECEMBER

We have had a strong gale of wind all day, the wind about north, and a most dreadful frost. It is really past bearing. We have driven a long way off the land today and the ice appears to be opening all around us. We are still driving the same way, being frozen up in the middle of a floe. We got the sun today. Her altitude was only two degrees—lat. 64°44′N. She rose exactly at half-past ten and set at half-past one. The way that we are driving to the southward will always help to keep her in sight the longer. We are on a very small allowance just now. I am really starving and have nothing to eat. There are two days in the week we get nothing to eat but one biscuit and a half. It will very soon bring me down. I feel my strength declining daily.... The cold tonight is really piercing — I never felt the like of it. I wish very much we had a thermometer.

SUNDAY 13 DECEMBER

This has been an extremely cold day, wind variable from NE to NNE and the weather cloudy. We are abreast today of a very remarkable headland. We suppose it to be Saunderson's Tower of Davis. It is of a most surprising height and resembles a fort and indeed you would almost think it was reared by human agency, but it was put there by a higher power. Nature in all its grandeur is the work of a nobler hand, even that same God that made man in his own image. We have had twice sermon today.... All hands were present both times. It is really pleasant to see how eager they are for the service and I have not the least hesitation in saying that the good effects of our meetings are taking root amongst us by making us all more agreeable and laying aside swearing, which I have not heard aboard the ship these many weeks. Our invalids are all about the same. There is one of them very bad today. Alas he has been spitting blood. Gravel [gallstones] is his complaint.

MONDAY 14 DECEMBER

...It was in 68°30′N we got first beset, which is a distance of 240 miles we have driven with the ice, not counting the westing, which would make upwards of 300 miles. The wind has freshened tonight again from the SSE, a very bad wind for us for driving us in upon the land. Saunderson's Tower bore per compass at 2 P.M. NE by E of us. The aurora appears very brilliant tonight. They are of a dun colour approaching to yellow. The bearings are about east. They appear to have a curious sort of tremulous motion. This is really a fearsome looking night. There is a very black cloud as dark as midnight hanging away to the NW which makes the night look so dismal, and the appearance of the aurora.

TUESDAY 15 DECEMBER

This has been another solemn, eventful day occasioned by the removal of another of our company to the world of spirits. He died this morning at three o'clock (he belonged to the *Middleton*). The time the soul left its earthly mansion it blew one of the fiercest gales that we have encountered. It was a hurricane with a blinding snow drift. You thought the very masts would be swept away but, thanks be to the Lord for his wonderful mercies, we lay as unharmed as we were lying in Dysart Dock [near Kirkcaldy].... We committed the lifeless body to the fathomless deep at 10 A.M. We had first to saw a hole in the ice, it was so thick. It was a mournful burial and dreadful as the angry blast swept through the confused and icy tackling, the frost so severe that you could hardly stand without pain on the ice for so short a time as burying. We all met in the half-deck after his body was launched into the deep and performed divine service [and]...tonight again at 6 P.M. for prayer and praise. The text was "Christ the anchor of the soul," a really impressive, sublime sermon for us and was very deeply received into the hearts of us all. We are now in the mouth of Cumberland Straits [Cumberland Sound]. I can see Cape Enderby and some islands right astern of us which bear WNW of us. We have had this gale from about ESE but it has moderated tonight. The aurora appears very brilliant tonight in all the different colours you can imagine. The *Jane* is out of sight now. We do know how she bears of us now. If any thing happens now with us it will be a most awful affair but we must alone look up to our Heavenly Father to pity and help us poor, miserable, blind, helpless, creatures through the name of his ever blessed and glorious son Jesus Christ, who died that we might live.

WEDNESDAY 16 DECEMBER

We have had better weather today but the frost nowise mitigated, the wind inclining mostly off the land. There happened to break open a lane of water about eight ship's lengths astern of us today. There were a great many seals and unicorns swimming about in it. We launched a boat over the ice to the hole to try and shoot some of the seals but I never got within shot of any of them. I wish much we could get a seal or two. We could eat them just now without hesitation....

THURSDAY 17 DECEMBER

...The lane of water that we launched the boat yesterday into, closed this morning with a frightful noise and would have torn the stoutest ship to pieces, yet we lay quiet and unharmed. It is almost miraculous when we look back upon what we have encountered, the many hairbreadth escapes, so many foreseen dangers that we could not avoid. Poor helpless mortals. If our Heavenly Father had not preserved us, where would we have been?...We will not lose the sun now which is a great benefit to us. We have eighteen hours darkness just now which makes a long, long, tedious night. It has come on tonight at 10 P.M. a strong gale from about north, a very fine wind for us for setting us a bit further from the land. The poor lad that got his feet frozen is still very bad. His feet are falling off by the ankles, actually rotted off, but the putrefraction has gone no farther up than the ankle. He is very weak. It would need a man with a strong constitution to be able to stand the shock of his feet falling off.

FRIDAY 18 DECEMBER

We have had a tremendous gale of wind this day from about north to N by E. We got the sun's altitude today again, which was two degrees and six miles [*sic*] and made our lat. 64°16′N. It is now about the shortest day and we are exactly 500 miles to the northward of home and 1800 miles west from it. At 8 P.M. it has come on one of the most awful gales that man ever witnessed. It is a hurricane. The ship being frozen up she gives nothing to the gale, which must strain her very much. You would think it would tear the mast out of her altogether. You cannot look at the deck. The cold is that piercing that you are not able to stand it above a few minutes. It has been my first watch tonight. At 11 P.M. I went up for a minute or two on deck and I saw a remarkable phenomenon [*aurora borealis*]. It was pitch dark. In a moment there was a bright, luminous arch shot around the sky to the NW from west, brighter than a hundred moons. I could have seen to have lifted a pen off the deck. It was in the form of a rainbow and lasted about four minutes, when it disappeared and left nothing but darkness and gloom. It was really grand, and a most imposing sight. I never saw any thing equal to it, indeed just now when we have neither sun or moon (but only the sun for a short time). It would be a prolonged gloomy, tedious and frightful, darkness if it were not for the thousand luminous streaks that every now and then shoot through the wintry sky, and so bright they are sometimes that for the space of a minute or so you would think that there were a dozen moons had shot forth their lights in the brightest effulgence. The ship has still lain very quiet, no pressure being amongst the ice tonight.

SATURDAY 19 DECEMBER

It has still blown a complete hurricane all this day....There were seven icebergs all in a straight line. By appearance it must have been a shoal....However there were two of

Whaleboats were light enough to be dragged over the ice to open water on their keels or on sleds when necessary. (National Library of Scotland)

them, tremendous masses, of a most surprising height. They made us very uneasy all day as they were right down on our drift and if we had come against any of them it would have been instant destruction. It is now 10 P.M. The gale has moderated a little but how far we are yet from them I can't tell for when it became dark we were still driving down upon them. O what an awful situation is ours. I may be writing these few lines and perhaps destruction is not 600 paces from our dwelling. It is terribly dark. There are one or two stars shooting forth but that is the most. How much need we have to keep our lamp trimmed, for we know not the hour that the Son of Man will come. Perhaps this moment the archangel of God is now warning us "that time shall be no longer." It is as likely as not (to our appearance) that the next moment our souls will be required of us and that solemn thought will

steal into the heart of the sinner when he sees nothing but an awful eternity before him. Am I prepared to meet my God? Am I prepared to die? Dreadful thought, indeed, to meet an angry God, and to be condemned with the awful sentence "depart from me ye cursed into everlasting fire, prepared for the devil and his angels." We met in the half-deck tonight again for prayer. I read a sermon to them—the text was in Jonah chapter 1, verses 4, 5—which had a great weight with us all. The subject was that if the mariner's trust was in God, no fears, no dangers of shipwreck, no affliction, would make his spirit fail.

20 DECEMBER to 4 JANUARY

If ever man experienced more afflictions and anxieties in the transitory world it has been us. The danger and the sights we have been subject to these last eleven days, no pen could write upon, but God was with us, and He and He only preserved us. My mind has been too much troubled for to have written any these few days past. We had twice to run and leave the ship for our life. The first time we left her was on the twenty-first of December. We drove down on the face of a tremendous iceberg. We were only within one boat's length of it when we shoved round it. The ice was turning up on the face of it about forty feet high with a thunder that would have made any heart shudder. What a sight to see us running away from her upon the ice with anything that we could muster of clothes, every step you took going up to the neck in snow, and the cold that severe that two hours would have been all we could have survived. O dreadful thought, when I recall the scene back into my memory, which with the divine blessing I hope, never, never, to forget. I saw God's wonders manifested to us in a manner that I do not think ever it was so seen before. Death within an oar's length of you. It is a solemn and striking thought indeed. The next time was on the twenty-third at two in the morning, pitch dark. A pressure took the ship (it was blowing a gale of wind from the SE at the same time) and lifted her clean on the top of the ice altogether. It was an awful scene. She eased again and fell down in about half an hour and to add to our dismal situation she was stove and from that day to this we have never left the pumps a moment. We have got a sail under her bottom but she is not much tighter. O, we cannot be too thankful for the Almighty preserving and watchful care over us.

...Yesterday God was pleased to call another of our little company away from us. It was the poor man that got his feet frozen. He must have suffered a great deal. His feet rotted off close by the ankles before he died. We buried him yesterday after performing funeral service. We met twice in the cabin yesterday for prayer and praise to almighty God in still preserving us in the land of the living. The text was Psalms chapter 95, verse 5 – "The sea is his and he made it." We have lain quiet all this day, the ice slackening away into lanes. We see the land to the north of Frobisher's Straits [Frobisher Bay]. Hawkes Island – we are about thirty miles from it.

5 JANUARY

We have had a tremendous gale of wind today from the WNW and hazy weather, the frost worse than we have had it at all. A true saying, "as the day lengthens the cold strengthens." We are still constantly at the pumps. We pump four minutes and stand fifteen minutes. We set the watches to the period of three hours a watch, which makes it a great deal easier, being shorter spells. The ship has lain very quiet although at times the gusts of wind you would think would tear her to pieces. The water we have pumped out of the ship is frozen to the amazing height of ten feet within these eight days – it just getting the same as a berg – but we have got a drain cut for the water to run away from the ship, which is a good job.

6 JANUARY

We have had better weather today but the gale still continuing. The ship has still remained very quiet. Watch still employed constantly at the pumps. I have been complaining these eight days with a very heavy cold, sore breast, and a great pain to draw my breath but almighty God, in whose hands I am, has been pleased to make me again whole. O thou my soul bless the Lord for all his benefits and be stirred up to bless and magnify his name. There are a great many of our men complaining, mostly of scurvy. Some of them are delirious. O it's a terrible sight to see them wasting away to shadows. Hunger is an awful thing. Yesterday I witnessed a scene that baffled description. We got a cask of blubber out of the *Jane* of Hull three months ago and we boil the blubber into kettles for oil to our lamps. If I did not see them eating the fins after they were boiled, whole pieces about two or three pounds weight! The smell alone of itself was enough to sicken any person. But, it shows plainly when a human being has not subsistence, he throws off his own nature and takes one of a more savage and desperate man. When he feels hunger and has nothing to appease it he is entirely a different being. What before he would have counted as almost poison he now would [seize?] with the greatest avidity. His temper too gets entirely changed, always fretting and always going about looking for something, at the same time not knowing what he is seeking. But thanks be to our merciful God I am still blessed with a good constitution, and my temper I think I have more command over. . . .

The *Jane* of Hull we have never seen these three weeks. We are much afraid of her. There are eighty souls on board of her, but I would fain hope they are all well. The frost has not been so severe today. We can see the body of water quite plain today. It bears (*the nearest water*) about ENE. We had the sun four degrees and seven miles [*sic*] high the day before yesterday which is a very cheery prospect for us that have come out of almost total darkness.

Scurvy continued its subversive work, producing among all those affected—at least a third of the complement—the horrid symptoms so familiar to students of arctic exploration. Several men had spent two months in their bunks and were incapable of moving their limbs. "Their gums are hanging away from their teeth altogether," wrote Elder. With each passing day every man on board the vessel found it a little harder to carry out any sort of work or exercise. They were on a starvation diet that lacked antiscorbutics, and unless fresh meat could be obtained to provide vitamin C they all faced the dismal prospect of steady physical deterioration towards death. As more men succumbed to the combined effects of scurvy, hunger, and exposure, gave up trying to keep active, and took to their bunks, the responsibility of carrying out vital shipboard duties fell upon the shoulders of a small number of comparatively able-bodied men—an additional burden at a time when their normal share of the ship's work must have seemed burden enough.

Gradually losing strength, those who were still on their feet became less inclined to make the effort to hunt, and less effective while doing it; thus the chances of securing the seals, bears, and foxes that held the key to everyone's survival diminished day by day. One fox had been caught a few months earlier and served up for New Year's Day dinner. A grand feast it must have been—some hardtack and salt meat, probably, along with the *pièce de résistance*, one miserable arctic fox spread among seventy-odd men! Or was it one fox shared exclusively by the captain and officers? Elder's journal fails to reveal how animals shot by the crew were divided. Did the hunter get his bag? Was it shared equally by everyone? Did it go to the sick, who needed it most? Or did it grace only the table of the master, mates, and harpooners? (Elder remarks on the fact that none of the cabin occupants was affected by scurvy.) In any case, William Elder was privileged to partake of the New Year's Day fox, and he found it "as tender as chicken." But was it simply his desperate hunger, he wondered, that made it so palatable?

Two more foxes and several small seabirds were taken in early January, but that was all; no other animal kills are recorded. The scurvy patients were encouraged to take exercise (which was already beyond the capabilities of many), and they were fed with salt water. This treatment probably did more harm than good.

Ironically, the means of combatting starvation and scurvy lay close at hand, but it was rejected. Although the *Viewforth* had not captured any whales before being beset in the autumn of 1835, there was on board a large cask of blubber, that had been obtained from the *Jane* in October. During the winter months small quantities of oil were rendered from this blubber to serve as fuel for the men's lamps, but it was not once used as food. Yet blubber constitutes a valuable caloric resource, which contains certain vitamins as well. And if the blubber had not been made-off—stripped of the skin and stringy tissue—the crew had at their disposal a quantity of whale skin, rich in vitamin C, a very effective preventative and cure for scurvy.

European whalemen were seldom inclined to eat whale skin and blubber. But hunger can be an effective antidote to food prejudice, and the crew of the *Viewforth* were gradually losing any qualms they may previously have had about such unorthodox dishes. Elder saw them secretly devouring three-pound chunks of blubber from which the oil had run out, and one man "begged for any sake that I would do him the greatest favour ever he got in his life if I would get a fine of blubber to him." But Elder, and doubtless the other officers as well, refused to let the men eat blubber and skin. They considered it a disgraceful habit, symptomatic of the savagery and desperation created by hunger, and took it as their duty to ensure that the men could not satisfy their base appetites on these loathsome items.

Unable to secure wild animals and forbidden to consume whale skin and blubber, the men had no alternative but to rely entirely upon the inadequate ship's stores. Four hundred pounds of bread had been salvaged from the wreck, but this represented no more than a two-week supply. The coffee, tea, and sugar issued weekly to each man appear to have run out by mid-December. In January the men were receiving two pounds of bread, or hardtack, per week – one biscuit a day – and if they were strong enough to be working at the pumps they were allotted an extra pound a week. Elder does not say whether they were still getting the half-pound of salt meat they had been receiving in late November; it seems unlikely.

Hunger made them desperate, and Elder's entries for the seventh and eighth of January reveal that the desperation was taking a more sinister form than the mere eating of blubber. These two entries have been boldly struck out in ink, presumably expunged by the author himself before he showed the journal to people in Scotland and finally donated it to his mother, but it is possible to make out most of the words. On 7 January the men evidently came aft and took the captain's brandy and gin, an act that Elder considered "nothing less than mutiny." On the next day he wrote that many of the crew "were in a mutinous state" and "if we are spared to get home they should be punished accordingly." What other specific actions there were is not explained, but the festering discontent does

not appear to have resulted in anything as serious as a takeover of the ship.

By 20 January the *Viewforth* was abreast of the northern tip of Labrador, having drifted southward about 600 miles in the pack ice. No one really expected that the ship would be released for a few more months, but fortune turned at last in their favour, for on the last day of the month the pack discharged the vessel onto the open waters of the North Atlantic. "We are now on the dark blue sea, escaped from the very jaws of destruction," Elder wrote in jubilation. Wearily but with renewed spirits the few seamen who were still capable of going aloft set sail upon the frosted yards and turned the ship towards Scotland.

Hopes were renewed but the battle was not over, for the ship had to be handled skilfully in the stormy, winter seas, and frequent sail alterations would be necessary. The crew was divided into two watches alternating on four-hour shifts, but at least thirty men were bed-ridden and many others were incapable of performing work. In Elder's watch only eight men were of any use and some of these had to "haul themselves along by their arms." The food ration was increased to six pounds of bread per week and a half-pound of salt beef per day for each man, but scurvy remained their constant companion. The death toll by 4 February was ten; twelve more were close to death, their time running out. Another man died on the seventh. On that day, a Sunday, the survivors met together below decks, "the sick people lying in their beds," and with William Elder's guidance they all gave thanks to God for their deliverance from the winter ordeal in the ice. "I thought it made a good deal of impression amongst them," confided Elder to his journal, and in view of the circumstances it is easy to believe that it did.

The *Viewforth* arrived in Stromness on 14 February with fourteen dead and only seven men able to work the vessel. Her earlier companion, the *Jane*, had come in the day before, having lost none of her crew, and another ice-trapped whaler, the *Abram*, had returned on 9 February with only one man dead after drifting down the Labrador coast halfway to Newfoundland. The last to arrive was the *Lady Jane*, with twenty-two dead.

A COPY OF VERSES ON THE
Dreadful Hardships and Privations

Of the unfortunate Seamen who are now confined with their ships, by the ice, in the desolate regions of Davis' Straits, where they proceeded early in the spring of 1835, with only 8 months' provisions on board, & are now living in that dark and stormy climate upon one biscuit per day,—May God prosper Captain Ross and his brave companions in their efforts to mitigate the distress of the sufferers, and grant they may obtain a speedy release and safe return to their disconsolate families and friends.

M. Montgomery, Printer, 26. Lowgate, HULL.

COME all friends of humanity, parents, sister
    or brother,
Who have got hearts within your breasts, to
    feel for one another;
Come list a while unto this tale of woe which
    I relate,
Concerning of those Seamen bold froze up in
    Davis' Straits.

'Twas early in the Spring of Eighteen-hundred
    and Thirty-five,
To prepare for the Whale Fishery, each man
    his best did strive;
Their wives, sweethearts, and children too, as
    they beheld them sail,
Did cry, "God prosper Sailors bold, who go
    to catch the whale."

Those Whaling Ships are Noble Ships to sail
    upon the sea,
There was the Duncombe, Lady Jane, the
    Swan, the Harmony;
There is the Lee and Isabell, which will re-
    turn no more,
For they were crushed by fields of ice where
    raging billows roar.

Those Noble Ships with several more, were
    manned with crews as brave
As ever ploughed the ocean wide or stemmed
    the briny wave;
With cheerful hearts they did depart, but little
    did they know,
What hardships great in Davis' Straits they'd
    have to undergo.

The Duncombe and the Harmony, thank heaven
    they did return;

While five hundred Seamen brave ice-bound
    in Davis' Straits do mourn,
To pass the dreary winter 'midst the northern
    cold severe.
'Midst bears and seals, and icy fields, where
    daylight seldom appears,

Now 'midst those dreary regions, from their
    friends far away,
Hunger and cold they do endure, with one
    biscuit a day;
To hear of hardships such as these each tender
    heart is grieved—
Kind heaven grant from their distress they soon
    may be relieved.

May Captain Ross and his brave crew soon
    reach that icy shore,
And soon succeed in transmitting provisions
    from their store;
To those poor men 'twill ease their pain, and
    mitigate their doom,
And likewise glad the aching hearts of wives
    and friends at home.

May God above have pity on those brave un-
    fortunate men,
And grant that safe and sound they shortly may
    return again;
And when in England they arrive from perils
    'midst the ice;
Then with their wives, sweethearts, & friends
    they will again rejoice.

So pity those poor Seamen brave who suffer
    hardships great,
With hunger and thirst, darkness and cold, ice-
    bound in Davis' Straits,

Handbill concerning the plight of "five hundred seamen brave ice-bound in Davis' Straits."
(Hull Museums)

A bad ice year: several whaleships beset in pack ice, four already crushed and their crews camped on the ice, salvaging material from the wrecks and driving off marauding bears. Sketch by John Gowland. (Dundee Museums)

Six Each Day the Frost More Bitter

All seemed now more than ever sensible of the necessity of early preparation for eternity.
DAVID GIBB, 1837

By 1836 SEVERAL THINGS had been made quite clear. The ice of Davis Strait and Baffin Bay could vary considerably in extent, duration, and behaviour from year to year. Its aberrations were wholly unpredictable, and calamities could occur on the West Side as well as along the Greenland coast. The ships were not adequately prepared for wintering; they contained insufficient provisions, insulation, heating devices, and fuel. Their crews were neither trained nor equipped for arctic hunting, travel, and survival. Those who had placed full confidence in the intrepidity, resolution, and perseverance of British seamen had been mistaken: these qualities by themselves were not enough.

Yet, incredibly, no substantial changes were wrought in the customary manner of dispatching whaleships to the northern fisheries. Was it reluctance on the part of the owners to invest money against the possibility of events that happened rarely and seemed unlikely to recur soon? Was it lack of compassion for the unfortunate sailors? Was it unreasoned hope that the Admiralty would through some ingenious plan assume responsibility for protecting whalers from the adverse forces of nature? Whatever the explanation, the English and Scottish ships set off as usual in 1836, their numbers reduced from sixty-nine to fifty-one. If their crews looked forward to a change from the unusually harsh ice conditions of the previous season or to an improvement in the outfitting of the vessels they were to be sadly disppointed.

Once again disaster overtook the Davis Strait fleet. Two vessels were crushed and another six were beset throughout the winter, repeating the awful adventures of the *Dundee* and the *Viewforth*. In the following spring the surviving vessels limped back to Britain with crews decimated by scurvy, starvation, and exposure, another shocking but little-heeded demonstration of the hazards of a winter drift.

One of the unfortunate vessels was the Aberdeen whaler *Dee*. She had set out on 2 April 1836 with a crew of thirty-three officers and men and had taken on an additional sixteen men at Stromness. To the complement of forty-nine were later added a number of survivors from the wrecked *Thomas*. But when the *Dee* returned to Scotland a year later she had only a dozen men alive, only three of whom were capable of going aloft.

After the voyage an anonymous editor who identified himself only as "J.H.W." compiled a book entitled *A narrative of the sufferings of the crew of the Dee, while beset at the ice at Davis' straits, during the winter of* 1836 *with other interesting and important particulars, drawn up by notes, taken at the time, by one of the seamen on board.* The seaman, evidently, was David Gibb. The narrative below is taken from this book published in Aberdeen in 1837.

The *Dee* did not reach Pond Inlet until 12 August but there were still plenty of whales; in the remainder of the month they killed four and picked up three drift whales to flense. When Captain Gamblin turned his vessel homeward in mid-September he found his way south blocked by ice. Another captain reported the Middle Ice of Baffin Bay equally impenetrable, so Gamblin resolved to try a northern route around the Middle Ice to the Greenland coast in company with the *Norfolk* and the *Grenville Bay*. The ships reached 75°N but finding no passage, returned to the Baffin Island coast, where they fell in with the *Advice* and the *Thomas* (on which John Wanless had served as surgeon two years before: see chapter 4). Wherever the ships turned the way was blocked.

There was now no prospect of getting out of Baffin Bay before winter. New ice was forming quickly on open water among the floes and although it enabled the men to walk from one ship to another it prevented the captains from working their vessels inshore to seek winter quarters in a protected anchorage. The five ships were prisoners of the pack, which steadily consolidated around them.

On 13 September the crew of the *Dee* agreed to a reduction of weekly bread allowance to three pounds per man "and a corresponding reduction in beef, meal, barley, etc." On the 27th Captain Gamblin reduced the ration further, to three pounds of bread and proportionately less of other foods.

From this date [10 October], the peculiar sufferings of the crew may be set down. There was not any farther reduction of allowances; but the fires having been extinguished from 8 P.M. till 5.30 A.M., the beds got exceedingly damp and uncomfortable; indeed, so much so, that there was a constant dropping in every one of them.

The chief points of consideration now, were the health of the crew, and how to keep every one in as active a state as possible. For this end, a variety of work was performed, which, in other circumstances, would have been perfectly useless. There were some things done, however, which, on account of the peculiar associations they called up, are worthy of particular notice. Such, for instance, as the sending down of the topgallant yards, unbending several of the sails, and unshipping the rudder. Seamen who have been accustomed to these duties know well that they are generally the harbingers to a safe and comfortable wintering; and, under such a prospect, it is pleasing to discharge them. But, alas! how different were the feelings of the seamen on board the *Dee* on this occasion! They had no pleasing hope of spending a happy winter with those who were near and dear to them—no joyful prospects of returning spring—no home—no comfort—no delightful enjoyment to anticipate at their own fireside. Instead of this, while the sails were unbending, and the halliards unwillingly yielding to the unreeving hand, the tears of sorrow and regret were mingled with the sighs of forlorn and almost desperate hope.

But the time was very soon to be otherwise occupied. The ice, as has been already mentioned, being so loose about the *Dee*, a fatal squeeze was hourly anticipated. This state of anxiety led to almost constant watchfulness, and the frost having become very hard, exposure was the more to be dreaded. On the fifteenth, the sun was again taken, and the latitude ascertained to be 72°58′N. On the sixteenth, it was 72°50′N — wind strong at NE, and large icebergs floating past. About this time the ice began to press hard, and was particularly severe, end on. During the night of the sixteenth, the vessel was hung by the quarter, the ice squeezing all along as high as the guard-boards. At daylight, the captain called all hands on deck, and ordered every one to lose no time in getting up the provisions. At 8 P.M. the wind fell off, the ship still hung up by the quarter. The ice, however, was quiet, and the prospects being rather more satisfactory as night fell down, several of the crew returned to take what rest they could get. At 11 P.M. there was another dreadful crash, but it passed over with less fearful consequences than were at first anticipated. On the eighteenth the ice gave way in several places, and opened up so far that a warp had to be got out to secure the *Dee*. All this time the other vessels lay undisturbed. Fresh water being much wanted, a boat was dragged across the ice to the nearest berg, distant three miles, and such duty was repeated frequently, and had a very baneful effect on the health of those who were thus employed. On the twentieth, the ice closed again with some sharp nips. To strengthen the ship, ten extra beams were put in aft, and the casks stowed in such a manner as to assist also in resisting the pressure. The strengthening was very seasonable, for the very next morning a very dangerous crash was felt; it ranged both fore and aft, and, for the time, led every one to think that the vessel was gone. It passed away, but only to give place to another in less than half an hour. This was a still more dreadful squeeze, and, under the impression that all was over, bags, chests, and every thing that would lift, were in one moment on the ice.

The dreadful sufferings which were this night experienced cannot be described. To enable the reader to form some idea of the situation of the men during this period, let him imagine one field of ice, of almost immeasurable extent, studded here and there with icebergs towering to the clouds. In a small spot is fixed the *Dee*, now reeling to this side and now to that, and every alternate roll attended with a crash, the sound of which more resembled the convulsive groans of an opening earthquake than the natural dashing of displaced water. On both sides, the crew are ranged at sufficient distance to avoid danger from the falling of the masts, and without any shelter, or fire, or other protection from the freezing elements of nature,

now severe in the extreme; and this, too, during the solitary hours of night, and without that comforting hope which is engendered by the prospect of an early dawn.

Such, then, were the circumstances in which the crew of the *Dee* were placed on the twentieth of October last, and human nature under these could not fail to quail before them. Contrary, however, to all expectation, the *Dee* was not injured; but, fearing another crash, the crew resolved to remain on the ice for a day or two more, now and then going aboard as the vessel eased. On the twenty-second, an observation was taken, and the latitude set down at 72°11′N. At ten at night, the ice broke up and drove, till about two next morning. Some dreadful crashes were then felt, and under the belief that no security was at hand, as much additional provisions as could be at all laid hold of were taken on the ice. While thus engaged, the most threatening squeeze yet experienced racked the *Dee* fore and aft. For safety, every man fled as he best could, and again under the dread of almost hopeless escape from the tottering masts. By and by the ice fell quiet, and as much of the provisions, &c. as possible, were again put on board. The pumps were tried, but no water was in the vessel. On the morning of the twenty-third, a good many lanes of water broke out; and here again the comparative comfort which had been for a few days enjoyed was disturbed; for it is invariably the case that, when these lanes make their appearance, the ice becomes very unsafe. Under the impression that another squeeze was pending, all hands were ordered to go on the ice with saws, and cut it up in as small pieces as possible. This was done, and some pieces being parbuckled and others sunk, the vessel was cleared a few feet. The *Dee* was farther eased at this time by the unexpected opening of a floe on the starboard side, and as soon after as possible she was hove a length in head. For a few hours, the greater part of the crew went to bed; but, at three in the morning of the twenty-fourth, she was squeezed again, by the one floe overlapping the other, and raising the vessel some two or three feet. Next morning,

The method of sawing through several feet of ice is shown in this sketch, drawn by an American whaleman in Hudson Bay. The weight of the blade helped to carry it down, after which the men hoisted it again with the tackle. From the Orray Taft *sketchbook. (Kendall Whaling Museum)*

the ice took off a little, and, as it appeared to be thinner right astern, the *Dee* was hove down about a hundred yards. What of the boats, chests, bedding, &c. as were not got on board at the former berth had to be dragged along on the ice to where she was now moored, but the difficulties in this duty can only be known by those who have had them to encounter. Every thing being again on board, and the ice getting more settled, hopes were entertained that the vessel would lie in safety until the spring.

[On the] twenty-fifth…Captain Gamblin here resolved to make every exertion to cut a dock for the *Dee*. With this view, several men from the *Grenville Bay* lent their willing assistance, and, with the aid of ice-saws and other means, a dock was cut out. This operation, however, cannot be passed by without specially noticing its nature and effects on the men. First of all, a triangle had to be raised on the

ice, and a block fixed in the junction at the top. Through this block a rope is reeved, to which an ice-saw is suspended, and from the other end bell or branch ropes, to allow five or six men to work the saws, and which is done in the same way piles are driven for laying foundation of guys or other buildings. The saw being raised as high as may be necessary, it is permitted to fall, and its own weight cuts downwards. When a piece of ice, large enough to be sunk, is cut out, as many of the men as can stand on the edge next the solid floe, weigh it down, and when the surface edge underlaps the floe, it is pulled in and kept down by the pressure of the heavier ice above. The great danger in this operation consists in the exposure of the men to cold by standing in the water, for it not unfrequently happens that the ice must be sunk so far as to permit the water to enter in by the top of the men's boots.

Several of the crew of the *Dee* felt the effects of this in having their feet frosted and their limbs rendered almost helpless. But to return, the frost was coming every day more severe, and so powerful, that the 'tween decks were all white as snow, and near where the fire had been through the day, icicles were found during night....

On the morning of the twenty-sixth [October], a solitary bear was seen; but just at the time when the crew were expecting to be benefited by such fresh provisions he made off,... The next three days passed away much more comfortably than could have been expected – the weather having moderated and the ice kept firm. During this period, the men were chiefly employed in carrying ice from the bergs to dissolve into water but, unfortunately, while one of the boats was being dragged across, the keel broke, and thus created additional labour and fatigue. Thus far the men were in good health, and under the prospect of being able to repeat the visits to the icebergs, a sledge was made, but, alas! never used for that purpose, as will be seen by and by.

On the twenty-ninth and thirtieth there was a change of weather for the worse; but the ice did not rise so much as was expected. On the latter day, three bears were seen and fired at, but without effect. This was trying, as fresh provisions were much wanted but there was one pleasing circumstance in connexion with this disappointment which is worth notice; two men having occasion to be on the ice perceived the bears, but fortunately made their escape in time to save their lives. It was but a respite, however, for the poor fellows, soon after, fell victims to the unrelenting hand of death, though laid on with milder severity.

November opened under rather more comfortable prospects than its predecessor exhibited. There was now no hope whatever of relief before the spring; and the ice being much more firm, the men had less harrassing duties to discharge. But even with these mitigating circumstances, what can we say of comfort in a latitude of 72°50′N – forty-nine men on a miserable pittance of provisions – beds and blankets beginning to freeze with ice, and little or no fire to dispel the cold. On the second of this month, a sad reverse was experienced, in the dock giving way in several places, and the ice again threatening to crush the vessel. Captain Gamblin, however, resolved on cutting a new dock; and, having again procured assistance from the *Grenville Bay*, and also from the *Norfolk*, forthwith set to work.... On the third, fourth, and fifth, the weather became exceedingly boisterous, and more snow fell than during any corresponding period since the vessels were beset; and what made the situation of the men more painful, the supply of coals was now nearly exhausted. In the meantime, one of the boats was broken up, and soon after another shared the same fate. On the sixth, the dock gave indications of rending; and, to provide against so fearful and so frequently fatal an incident, all hands were ordered to cut the ice in end as far as thirty fathoms. This was done; but the frost became so intense that the first length closed before the vessel could be hove in. This day another observation was taken, and the latitude found to be 72°23′N. In the afternoon, the wind veered from SW to SE, and towards night the ice opened in all directions. Serious fears were again entertained for the safety of the vessel; and, at night, so sharp were the nips, that more than once all hands were prepared with the bags again to go on the ice. At eleven o'clock, destruction seemed inevitable; but, soon after lowering the boat and leaving the vessel, the ice immediately closed, and thus permitted the men again to go on board. The provisions were farther reduced on the seventh, the heaviest reductions falling on the beef and barley. On the eighth and ninth, the ice was very unsettled, and little or no rest could be got for fear of a squeeze. Several of the men were severely frostbit in the face at this time, and relief was only got by rubbing with snow the parts affected. In the afternoon, two bears were seen, but like the others made their escape. The morning of the tenth looked well; and Captain Gamblin, thinking a safer dock might be cut, assistance, as formerly, was procured from the other vessels. The dock being finished, the *Dee* was hove in, and the ice, to all appearance, was more secure than it ever appeared before. On the eleventh...two foxes were seen near the *Dee*, but they also made their escape.

By this time, however, several of the men had managed to cook the tails of the whales so as to form a tolerably agreeable meal. But the fact is, very few were now fastidious in their taste, and it might have been well had everyone endeavoured to accustom himself to the provision in question; those who did so felt themselves much benefited. For the advantage of others who may happen to be placed in the like distressing situation, it may be well to give a brief outline of the manner in which the tail was cooked. The detail, too, will be interesting to the general reader. In the first place, it was cut up into pieces about four inches square, then laid out on the ice to bleach, or be purified by the frost, where it lay for four days; next, par-boiled, then allowed to settle on the top; the top was then scummed off, the pieces lifted out and laid past till they were wanted; and, when about to be used, were fried in the pan with a little fat. Delicacy, it will thus be seen, was here out of the question; and, however exceptionable the doctrine of the "end justifying the means" may be, it will be obvious to every one that, in the present case, it was justifiable.

On the twelfth [November], the sun was almost gone, and the weather excessively cold. Captain Gamblin, in the forenoon, gave each man a yard of canvas to make into snow-boots with wooden soles, which proved of great benefit to the men. Had it not been for this, many more would have had their feet frostbitten. To allow flannel to be rolled round the feet, the boots were made very large, and the wooden soles were remarkably well adapted for keeping out the frost.

The thirteenth being Sunday, Mr Littlejohn, the surgeon, read two sermons and prayers; and the value of religion seemed more to be felt at this moment than ever it

had been before. The crew had been very anxious, for some time, to have religious exercise on board; but at first they were rather diffident in requesting Mr Littlejohn to take the lead. A deputation, however, waited on him, expressed their anxiety on this head, and it says much for that young gentleman that he readily and most cheerfully complied with the request; and not only did he lead the exercise on Sundays, but also on other days, more particularly on Wednesdays and Fridays. Here we must pause for a moment, and allow the reader to reflect on the solemnities of such a scene. Think, then, of forty-nine fellow beings assembled together in one common bond of Christian worship; but no gilded temple shone around the worshippers—no taught minister of salvation declaimed from the pulpit—no eloquent oration, the production of calmful and uninterrupted meditation, fell on the ears of the assembly—nor any church-going bell tolled the hour of dismissal. But what is worth more than all these, there was but one sentiment, one feeling, one motive, and one object in view; and this one sentiment and feeling and motive and object were the offspring of a sincere conviction that the realities of eternity were soon about to burst on them....

On the fifteenth, the sun did not make his appearance. This was a sorrowful day. Each one surveyed the heavens, alas! with anxious thought, and but too sensibly observed a "blank in nature." It was a change which none on board had ever experienced; and what rendered the contemplation more painful, was the very distant hope of ever seeing his brightness again....

December opened with gloomy prospects; the effects of so much exposure and harassing duties being now severely felt. The first indications of disease were coughs, swelled limbs, and general debility; and it is worthy of particular notice, that the flesh became discoloured with small red spots, attended with sharp pains and stiffness of the limbs. Depression of spirit invariably followed when these symptoms of disease were perceived, and even with the opportunity of active and healthful exercise there was

Arctic fox. (Roy Plant — National Film Board)

little or no inclination to profit therefrom. On the third, shot two foxes, but "what were they among so many?" Those, however, who did partake of them, expressed their confidence of having been much refreshed. On the fifth, another fox was killed, and the appearance of several more induced the hope that a supply of fresh provisions might yet be at hand. On the eighth, another observation was taken, lat. 71°12′N. At this time, the wind was easy, the sky clear, and the northern lights very brilliant; but were never heard so distinctly as on the Western Ocean.

The *Thomas* of Dundee was listed over on the twelfth by a heavy pressure of ice, and so far did she heel, that the crew were obliged to creep on their hands and knees over the deck. This situation, distressing as it would have been at any time, was now much more so, the frost being so severe and the daylight almost gone.

On the morning of the thirteenth, a melancholy spectacle presented itself. But a few hours before, the *Thomas*, though heeled, was not damaged; but now she lay a total wreck. The masts were not yet gone by the board, but the vessel was past recovery. About midday, twenty-two men went over from the *Dee* to the *Advice*, which lay between the *Dee* and the *Thomas*; but being unable to proceed farther that night, resolved to abide there until morning. The crew of the *Advice* at this time were pretty healthy; but the provisions were so short, that the men from the *Dee* could only get one half-pound of meal each man for supper, and the same next morning. At eight o'clock in the morning of fourteenth, about sixty men from the *Dee* and *Advice* went over to the *Thomas*, with a view to save as much of the provisions as possible, and render assistance to the men in saving their clothes, &c. Three days were spent in this hazardous work, and when all were collected, an equitable distribution among the whole of the vessels took place, both of men and provisions. It is right to notice that two of the crew of the

Thomas died on the ice the night she was wrecked. These were the first deaths; and while the surviving could not help thinking that a similar fate, though perhaps under different circumstances, awaited themselves. While engaged in carrying the provisions from the *Thomas*, much injury was sustained through the cold. The ice, too, over which they had to pass, was exceedingly unequal, and, in different places, was divided by lanes of water. In many cases, the men sank down into the water, and thus laid the seeds of that disease which soon after proved so fatal. The supply of provisions which the proportion of the *Thomas*'s crew retained was much about the same as the mess for those of the *Dee*. The chief point of regret in regard to the loss of the *Thomas* was the distance; because that precluded the possibility of breaking her up for firewood.

A lunar observation was taken by Captain Gamblin on the sixteenth, and the latitude set down at 70°29′N.... About this time, Captain Gamblin suggested a farther reduction of allowance, but the crew would not on any account submit.

The death-monster scurvy now began to harass the greater part of the men. The disease first appeared in the mouth, and was known by a swelling of the gum and deadening pain, which increased to an excruciating feeling when anything touched the parts. When beef was used [eaten?], the suffering was dreadful, and the salt and cold frequently caused the blood to flow. On the eighteenth, twenty-one men were ill of scurvy, some of them suffering most severely. To add to the misery of all on board, the ice again gave way and threatened to squeeze every one of the vessels. The *Advice* seemed to suffer most; and, indeed, so dangerous was her situation at one time, that all hands were ready to spring on the ice, where the bags and a good deal of her provisions were already placed. The *Grenville Bay* was in great danger on the eigteenth, the ice having opened and squeezed as far up as the stern windows. Captain Taylor had almost despaired of her safety, as was indicated by a light on the bow, the concerted signal of

danger. A great deal of snow fell on the twenty-third. The twenty-fourth is remembered chiefly on account of another reduction of the allowance. The bread was not reduced, but the pork and beef were. The quantity weighed out was half a pound of pork each man per day; and, alternatively, with three quarters of a pound of beef. From this date to the end of the month, several of the crew were confined to bed by scurvy, and their situation there was anything but comfortable....

As THE YEAR DREW TO A CLOSE the four ships *Advice, Dee, Grenville Bay,* and *Norfolk* remained embedded in the pack ice, with the crew of the wrecked *Thomas* shared among them. It was doubtless some consolation that they were all within sight, each ready to provide manpower or shelter if danger threatened one vessel in the group. Nonetheless the situation was awesome. Because the ships had frozen in extraordinarily far north – abreast of Pond Inlet at 73°N – they were still about 70°N at the end of December, farther north than the *Viewforth, Jane,* and *Middleton* had been in the fall of 1835 at the very beginning of their drift voyages (chapter 5).

Mid-winter had passed, at least, and the men (like those of the ice-bound *Dundee* in 1826–27: chapter 2) looked forward anxiously to the return of the sun and to ameliorating temperatures. But January came in with piercing cold.

The first of January, 1837, was a day of sorrowful remembrance. Fain would the mind have banished the recollection of a New-Year's Day at home; but every effort seemed fruitless. Memory conjured up with avidity every pleasurable enjoyment, and seemed, as it were, to revel in delight when the conflicting remembrance of past and present experience clashed together; and, O how dismal the contrast! On that day twelvemonths, almost each one could boast of a home, and not a few enjoyed the endearments of social bliss. Now, the husband had no smiling partner to comfort him in the intricacies of life, to soothe him in his sorrows, to enliven his solitude with the balm of conversation, and render his home the soft green on which his mind would love to repose; and now, too, he had no little ones prattling at his knee, lisping their affectionate attachment in words of "softest tenderness," and gladdening his care-worn heart with their smiles of innocence and filial obedience. And again, the youth in "beauty's pride" had no fond one near on whom his best earthly hopes centred, and whose reciprocal and faithful and tenderly-devoted attachment was the pleasure of his life and the ardent delight of his hopes and prospects. Or it may have been, and indeed it was, that the dutiful son was now deprived of another marked period when the most valued testimony of his sincere regard would have been tendered to a doating father or mother. Instead of this, there was nothing but blackness and darkness around. Disease had fixed his fell grasp on his intended victims; the writhings of his torment pierced the very soul, and despairing hope almost changed the best feelings of nature into the fiendish passion of a maddened brain. And farther, the elements of nature combined together, as it were, in order to complete their misery. Aloft, the tempest waged a dismal warfare with the yielding and tottering masts, and the wind wildly screamed through the frozen shrouds as if the death whistle had already sought to disturb the peace of the mariners below. Nor was this all, the crashing of the contending bergs as they rushed against each other and burst asunder, seemed to shake the very ocean, and spoke like the yielding forest in the mastery of the gale....

The weather on the second, third, and fourth of this month was very moderate; but the frost still exceedingly hard. Scurvy was daily making rapid strides among the men, and seemed to threaten universal destruction. Mr Littlejohn the surgeon applied the most effective medicines he had, and continued his unwearied exertions with a degree of anxiety and care which does him the

highest honour. But the disease could only have been cured by fresh provisions; and, alas! the prospect of such a remedy was distant indeed.

It occurred to the men, on the fifth, that the quantity of provisions on board would allow some additional allowance, and, with this view, a deputation waited on the captain to make the request; but Captain Gamblin thought it would be premature to interfere with the present mess, and, therefore, declined to grant the request. He mentioned, at the same time, that he was quite satisfied his men had the power to compel him to comply with their wishes, but he would hope that they knew themselves better than to attempt such a line of conduct. He was right in his conjecture; nothing was farther from the intention of the seamen than to disobey their commander's instructions, and, therefore, they would rather starve than take by force what was not given of pleasure. These sentiments being expressed to Captain Gamblin, he reconsidered their application, and ventured to add a little additional flour.

The brilliancy of the sky on the morning of the sixth, gave hope that the sun was soon about to make his appearance, and as the little daylight there was increased in brightness, the watch were gratified with a view of a large sheet of water on the starboard quarter, distant only a few miles. Such a scene earlier in the season would have been viewed with dismay; but now the possibility of relief was, to say the least of it, more rational. These anticipations, however, were too good to be realized and too pleasing to be gratified.

On the watch being called on the morning of the seventh, a large proportion of the men were unable to leave their beds. This was a melancholy state of things; and what now made the case still more painful was the fact that the beds were in a most deplorable state of cold and also of vermin but more of this by and by....

Hitherto disease had dealt out his distressing pangs with torturing severity, but life had still been spared. "The great change," however, was near, and as the morning of the eleventh unwillingly lifted up its shaded light, one soul "returned to the God who gave it." This being the first death, it may reasonably be presumed that some account of the last scene will be expected. The deceased, William Curryall, a native of Stromness, was in his fiftieth year, apparently of a healthy constitution, but rather subject to nervous debility. For some considerable time he had been labouring under depression of spirit, and seemed to have given up all hope of relief from the day he was seized with scurvy. As the disease strengthened, his mind got rather weak, and up to the time of his death did not recover its wonted tranquillity. He died without a struggle. As soon as possible the body was carefully wound in a blanket, sewed up, laid out on the carpenter's bench, and then all were requested, who were able to assemble, to join in the last service. This call was willingly obeyed, and each one seemed to feel with deepest sincerity the full force of the solemnities of such a scene. The funeral prayer was read by Mr Littlejohn, and then the corpse was carried, in mournful procession, to an opening in the ice, through which it was consigned to a watery grave.

On the fourteenth [January], the daylight was getting pretty strong, and the sky kept very clear. About noon, several lanes of water were seen, and what was more rare some whales were blowing in them. These were the first that had been seen since the *Dee* was beset; but the men were unable to make any attempt to kill them, and besides, it would have been almost, if not altogether, impossible to have done so.

Nothing particular occurred until the sixteenth. On that day there was joy even in the midst of grief. By the calculations made from previous observations it was not expected that the sun would be seen before the twenty-fifth; but most unexpectedly he appeared on the sixteenth. This was indeed a pleasing sight, and so delighted was each one when he saw the "orb of day," that half his load of woes seemed to have been removed. ...

The sun's altitude had hitherto been taken on board the *Dee* by Captain Gamblin, but now he was unable to do so. For some weeks he had been complaining, and seemed to suffer, too, from depression of spirit. The utmost attention was paid to his wants, and everything done which could in any way relieve them, but he made no progress towards health. Under these trying circumstances, the mate took an observation, and set down the latitude at 69°01′N.

Seeing the crew getting weaker day by day, with no prospect of recovery, the mate thought it would be well to take in two reefs of the topsails. When this duty was ordered, only fifteen men were able to go aloft, and of these the greater part were very weak. The motive which induced the mate to reef the sails, was conviction of its being exceedingly probable that the effective hands would be so far reduced as to be unable to shorten sail should the vessel get out and a gale come on. [Presumably topsails had been set to assist the ship's progress and manoeuvrability within the pack whenever leads of water opened up].

William Besley, from Aberdeen, died on the nineteenth. The sufferings he underwent were of the most excruciating kind, but he seemed to endure all with patient resignation. The same services were gone through in this case as in the other, and his body was sunk at a considerable distance from the vessel; those who were able attending in mournful procession.

So soon as the announcement of another death reached the ears of the sick, each one gazed in silent thought on his fellow, but none expressed the feelings of his mind. It was often observed, however, that such announcements at first caused much depression of spirit; but, when the deaths had become more frequent, they were less impressive on the survivors.

The ice was now remaining firm, and the winds were light, but the frost was getting more and more distressing, with no appearance of a thaw. During the last week of this month [January] several whales were seen, some of them very near; unicorns were also observed in the lanes of water; but there was no one now caring about anything but his life, and even life itself seemed a burden. The melancholy tidings of the Captain's hopeless state had a severe effect on the men, and led several of those who were inclined to hope for life now to despair. It was a convincing proof of the mortality of the disease that, even with the privileges the Captain possessed, it could not be stayed; and surely, if in this case death was almost certain, there could be little hope between decks.

Andrew Bennett, from Aberdeen, died on the twenty-seventh. He had caught cold by the exposure; and, being of rather a delicate constitution, his lungs were injured. He was also seized with scurvy, which, together with difficulty of breathing, made his case one of peculiar distress. He was quite sensible to the last, and died very quietly.

January ended by a most disagreeable change of weather. The ice again threatened to break up and crush the vessel; but, by this time, many were so tormented with pain that death seemed a consummation rather to be wished than dreaded.

The month of February opened under the most alarming prospects. Three of the crew had already "gone the way of all the earth"; and the fatal disease was making such dreadful inroads on the health of the survivors, that a serious increase of mortality was hourly expected. John Setchell, a native of Hull, but married in Aberdeen, expired on the evening of the first. Throughout the voyage, Setchell had been weakly, and his predisposition made his death almost a matter of certainty, now that scurvy was so common. The second saw no blank; but, on the third Captain Gamblin was no more. This was the heaviest stroke, and could not fail to throw a gloom over the already dark and dismal scene. Captain Gamblin's sufferings were most excruciating, and more particularly

towards the close of his life. His body was wasted to a shadow, and his well-known manly and powerful voice could only breathe like the whisperings of a child. The nervous debility which had pressed heavily on him during the earlier stages of his sufferings increased to a painful extent a few days before he died; and the thought of his wife and family at home so distracted his mind, that at times he gave vent to the most frantic grief. Compared with this, his own suffering seemed to give him little trouble; and, if there was one motive which, more powerfully than any other, induced an anxious hope of relief, it was the earnest wish that he might be spared to his family. Under this impression, he was incessantly talking about home; and, even when the strength of disease had deprived his mind of its wonted composure, the burden of his soul was his dear wife and little ones. But his hour was come—he died; and, while dying, breathed a last and fervent prayer that a kind Providence would realize, in the experience of those whom he had left behind, the fulfilment of the promise, "I will be a father to the fatherless, and the orphan's stay."...It was the sixth hour, and all who were able assembled in the 'tween decks, and within hearing of those who were on the beds of death. Each one was silently and deeply impressed with the realities of the heart-rending scene, and...the mind was in a peculiar manner disposed to listen to the instructions of the sacred volume. Mr Littlejohn had just begun the service, a passage from the New Testament was being read; and, while all were attentive and earnest, the death-toll was heard—the spirit of the Captain had that moment fled.

From the peculiar intimacy which had subsisted between Captain Gamblin and Captain Taylor of the *Grenville Bay*, the mate despatched a message announcing the melancholy intelligence. Captain Taylor immediately came over to the *Dee*; and, on consulting with the mate and the surgeon, it was resolved to preserve the body of

Captain Gamblin as long as possible. A coffin was therefore made by the carpenter of the *Grenville Bay*, the carpenter of the *Dee* being confined to bed, and the body was then placed on the quarter-deck.

During the two following days, a gale of wind from the east again threatened the vessel, and the danger was the more increased on account of the helpless state of the crew. Every hour some one was taking to bed; and of those who were able to walk about, only some eight or nine could do duty. The drift south, in the course of these two days, was only about eleven miles; but, though the latitude was getting lower, the frost increased, and was much more severe than when they were as far north as 75°N. To give some idea of the intensity of the frost at this time, it may be stated that every liquid was frozen; the water in the casks was a piece of solid ice; and, even while the snow was being melted to cook the victuals, the icicles were hanging by the side of the scuttlelass, or water-cask, distant from the coppers or fire-place about six feet! Every chest between decks was white with frost, both outside and inside; and, as a proof of this, it need only be mentioned that, so soon as a thaw came, some pints of water were in every one of them. With such severe frost there, what must have been the sufferings of those who were not able to come within sight of the fire! How deplorable a sight was it to behold the very blankets in the bed covered with solid ice, especially by the sides of the vessel – the pillows frozen in every part but where the head lay, the very hairs of which were in some cases stiff with cold – each day the frost getting more bitter, and each hour the unfortunate sufferers becoming more and more unable even to break the ice on their wretched coverings. About this time, too, vermin began to make their appearance, and so suddenly did they increase, that in some of the beds they were literally swarming; and these vermin were of a most rapacious kind. In many cases, they found a lodgement underneath the skin, and fed on the flesh like cancerous lechers. But so horrifying

would the details in this case be, were they minutely given, that no human being would believe them. Besides scurvy and vermin, the sufferers were almost seized with most violent diarrhoea; and so dreadfully did it affect them, that relief for a single half hour would have been hailed as an invaluable respite. Under such a complication of disease, it was impossible for the mind to be otherwise than in a state of mental abstraction; and, while this prevailed generally, it was remarked that those in particular who were conscious of having lived a thoughtless life were in the most deplorable state; and, indeed, so awful were some of the cases, that the survivors say their expressions of despairing prospects can never be forgotten.

George Dawson, from Shields, a spectioneer, died on the fifth. In the afternoon, the mate made a calculation of the quantity of provisions, and the probable time of getting clear; and, believing that some additional allowance might be given, he decided on giving it on the bread and barley. This was matter of great thankfulness to those who were able to partake of food, and is believed to have in some measure tended to preserve life which would otherwise have been lost. The spirits had been reduced for some time; but the men generally did not give themselves much trouble about that, no good having been derived from them. But the severest hardship, apart from bodily pain, was the want of fire. The coals had been done since January, and stocks, or staves of casks, were all that could now be had. These were not sufficient to throw out heat so as to dispel the frost between decks; and, what was worse, they were only made use of to cook the victuals. The only other time fire was allowed, was when ice had to be melted for water to drink; and here it may be observed that, when the men were thirsty, they had to wait with patience until the snow was dissolved, then watch the cooling of the water, and seize the earliest chance of getting a draught before it was again frozen. Had coffee or sugar been plentiful, the water might have been used warm, but such was not the case. A few of the men had yet some coffee and tea, but the greater part had none.

On the three following days there were no deaths, but, on the tenth, Christopher Sutherland expired; and, on the eleventh William Flett. The former belonged to Shields, and the latter to the Orkneys; neither of them were married. On the twelfth, John Isbister, Robert Burns, and Alex. Reid, died. The boy Reid expired in the arms of David Gibb; but his death was more the effect of palpitation of the heart than scurvy. This boy died quite sensible and very happy....William Muir, one of the men of the *Thomas*, died on the fifteenth.

The West Land was seen on the sixteenth, in latitude 63°33′N; and the water appearing to get rather more free, strong hopes of getting clear soon were now entertained. The other vessels were getting now pretty far north. The *Advice* was as far as twenty miles distant. No communication had been held with the *Advice* since the end of December. At that time, two of the men came over to the *Dee*, and represented the state of the crew to be so very sickly, that no more on board were able to travel so far had they been inclined to do so....Scurvy was the disease from which the men were suffering.

No more deaths happened on board the *Dee* until the twenty-third. On that day, John Dunbar died. He belonged to Aberdeen. John Learmond died on the twenty-fourth. Samuel Brown, from Shields, the cook, and James Gaudie, both died on the twenty-fifth. On the twenty-sixth, John M'Leod died; and, on the twenty-seventh, John Booth also expired.

March revived peculiar recollections. In previous years, nearly all on board had been busily occupied during this month in making preparations for the fishing. *Then* they looked anxiously and hopefully forward; but, *now,* how different the prospect! There was, indeed, something peculiarly pleasing in anticipating a return home...but,

then, every day reduced the number of the crew; every hour, those who were as yet comparatively healthy were falling victims to disease; and the probability of no effective assistance being awaiting previous to reaching home—all conspired to darken the hope and dash the expectation.

Alexander Anderson died on the first of March. The wind prevailed from ENE during the next three days of the month, and the ice became very loose and dangerous. At one time, the probability of a fatal squeeze was so strong, that the provisions were all put in readiness for being taken on the ice; and, while these preparations were making, what must have been the feelings of those who were unable to be removed from their beds?

The ice closed on the third; and the foretopsail, which had been set in expectation of speedy relief, was again stowed. On this day, Robert Moir, of the *Thomas*, died. On the fourth, John Booth, from Aberdeen; and, on the sixth, Andrew Masson, also from Aberdeen, both of the *Dee*'s crew, shared the same fate. On the seventh, James Yorston expired; but no more were cut off during the next four days.

The latitude on the ninth was 62°49′N; wind NW blowing very strong. By this time, only six hands were able to do duty; and, under the impression that the *Grenville Bay*'s crew were more healthy, the mate of the *Dee* went over and requested to know if Captain Taylor could render any assistance were the vessel in the open sea. Captain Taylor did not consider that he would be justified in doing so, as twenty of his own men were on the sick list. About this time, full allowance was given in everything except the bread, which was limited to three and a half pounds per week, each man, but it was too late, few were able to eat it. The *Norfolk* was now about seven miles distant, bearing N by W. The *Advice* nearly as far off, bearing ENE.

On the eleventh, the *Advice* was seen with her sails set,

and making way through the ice towards the south. The latitude was this day ascertained to be 62°43′N. In the afternoon, Peter Linklater, of Orkney, died. On the fourteenth, James Muir expired, also a native of Orkney. A great deal of snow fell on the fifteenth; and at 2 A.M. a heavy swell came on, which broke up the ice. About this time, the vessel was in great danger, some of the ice almost breaking over her. At 4 P.M. the rudder was shipped, and strong expectations entertained of getting early out. William Stirling, of Aberdeen, died during the night, and was the last one whose corpse was sewed up in a blanket and put down through the ice.

The sixteenth was a dull frosty day, but the ice was breaking up in all directions, and gave other indications of a clear opening being at hand.

Five months and eight days had now elapsed since the *Dee* was beset in the ice, during which she had drifted 670 miles south; and after such dreadful and unprecedented sufferings, it may safely be presumed that the survivors were glad to get out. But, alas! many of those who had yet hope of seeing home were denied such happiness.

IN MARCH, the day *Dee* escaped from the ice, David Dinnet and James Cook died of scurvy. During the next five weeks, as the ship laboured homeward across the North Atlantic, weathering strong gales on three occasions, the deaths continued; John Moir; William Paterson; Magnus Curryall; Alexander Noble; Alexander Garden; James Pearson; James Tulloch; G. Turriff; Allan Monro; William Anderson; William Turriff; Andrew Spence; Thomas Stuger; John Davidson; David Smith; James Isbister; David Irvine—all dead from scurvy.

The first chance of aid came with the sighting of a brig on 20 April but when the men hoisted a signal of distress the unknown vessel continued on her course and disappeared, leaving the disconsolate crew of the *Dee* to return to the problems at hand—a leaking hull, sails that split, and ropes that parted. The weakened condition of the surviving men made these jobs and the routine business of trimming and changing sail enormously difficult.

Islands were sighted on the 24th and on the next day the assistance they so badly needed finally arrived.

25 APRIL

Moderate breezes, with clear weather. Saw a fishing-boat; hoisted a signal; boat came alongside; learned from the fishermen that we were nigh the Butt of Lewis. At 6 P.M. the barque *Washington* of Dundee, Mr Barnett, bound for New York, bore down upon us, and inquired if we wanted any assistance. On informing him that we had only three hands able to go aloft, he sent four men to our assistance; came on board himself, bringing with him wine, porter, &c.

26 APRIL

Light breezes, with clear weather. The barque took our ship in tow, and towed her to the sound; committed her to the care of a pilot. In the evening, the surgeon appointed by Government came on board to visit the sick. Anchored in Stromness harbour at 11 P.M.

27 APRIL

Light breezes, with clear weather. Ship at anchor in Stromness harbour. Early in the morning, the surgeon came on board, and got the sick conveyed on shore to the hospital. Several men sent on board to do sundry jobs.

The most remarkable incidents peculiar to the voyage from the ice to Stromness were the refusal of the vessel seen to render assistance, and of the crew in the boat to come on board [25 April]. It were premature to condemn the conduct of those on board the ship, as it may have been

THE MISSING WHALERS.

By the following letter, received yesterday, by Messrs. G. and J. Egginton and Sons, it will be perceived that one of long absent fishing ships, the Dee, of Aberdeen, has at length arrived at Stromness with only nine of her own crew, which consisted of forty-six, and three others, the survivors of twelve men belonging to the THOMAS, of Dundee, which was wrecked on the 13th of December last. Although the below named vessels were supposed to have been beset in the ice at the same time as the SWAN, their is no account of her.

STROMNESS, April 26th, 1837.

Messrs. EGGINTON AND SONS.

GENTLEMEN.—The bark Dee, of Aberdeen, arrived here this evening, from Davis Straits, with seven fish, about sixty tons. Got out of the ice on the 16th March, in latitude 62 deg. 15 min., together with the Greenville Bay and Norfolk, and they suppose the Advice would get out at the same time. Thirty seven of the Dee's crew have died and amongst the number is Captain Gambling. The Thomas, of Dundee, was lost on, the 13th of December; twelve of the crew were put on board the Dee; nine of them have since died, which makes forty-six in all that have died on board the unfortunate "Dee." Have heard nothing of the Swan, of Hull, and seen no ships on their passage home. The Washington of Dundee, fell in with the Dee, of the Butt of the Lewis, and towed her into Holy Sound,—the remainder of her crew (twelve) are very ill. There is another ship in the offing, which we suppose will be one of the other ships. I write this in haste, with a view to get it off per the steamer, Sovereign.

I remain, very respectfully,

Gentlemen,

Your obedient Servant,

JAMES DAVIDSON.

Weston Howe, Printer, 33, Lowgate, Hull.

Handbill announcing the return of the Dee *to Stromness after being trapped in the Davis Strait pack ice during the winter of 1836-37. (Hull Museums)* that they did not perceive the signals, but the men in the boat are not so excusable. The paragraph in the foregoing log, which refers to the boat coming alongside is not so explicit as the following, which was noted at the time by one of the seamen:

This morning, at six o'clock, we saw a boat containing eight men; we hove to, close to the windward of the boat; we waved upon them to come alongside; they came and asked us in English what we wanted. We told them that we had been all winter in Davis' Straits; and that, if they would come on board, and assist us to Stromness, they would be well paid for it. They would not come on board. We hove two pieces of meat into their boat; the boat being half loaded with fish, we hove down a rope to them, and asked of them to give us a fish, but they refused, and pushed off.

Now, we cannot conjecture what could have been the motives which induced these fishermen to refuse assistance; and not only so, but to forgo the prospect of a rich reward. Were it not that we know British seamen better than to believe that they could be callous or indifferent in such a case as this, we should condemn these men at once as a set of merciless scoundrels. But, when we think on the sickly appearance of the men on board the *Dee*, the coffin on the quarter-deck, and other forbidding circumstances, we are inclined to believe that the fishermen suspected a case of plague, and thus refused to put themselves in a situation where they would most likely fall victims to its dire influence. But the taking of the pieces of beef and refusing to give a fish in return, are drawbacks on this opinion; and, if the men can reconcile such conduct with the claims of suffering humanity, they are more objects of pity than of hatred. By order of the owners, Captain Goldie, from Aberdeen, went to Stromness, and, having engaged effective hands, brought the *Dee* to Aberdeen.

There is another circumstance, in connexion with the voyage from the ice, which is worthy of particular notice. The men who were spared were inclined to hope that some Government vessels would be cruising in search of the missing whalers; but, having fallen in with none, though sailing in such wide latitudes as the vessel did during the first few weeks after she got clear, the men got low in spirit, which, perhaps, tended to hasten the melancholy fate of those who so soon thereafter died.

The *Dee* was not so much damaged as might have been expected, and indeed, looked much cleaner than many vessels after an ordinary voyage. She got some cleaning at Stromness; but, even when she arrived there, she was not damaged to any extent worth mentioning. [This]...can only be accounted for in the great strength of the vessel.

A good deal of provisions were brought home; and some persons have been uncharitable enough to throw blame on certain parties, whose conduct, they think, in reducing the allowance to one-half so early in the season, is questionable. This is a rash condemnation; it need only be known that the *Dee* was very early beset, and in a latitude where there was every probability of a long wintering, to do away with any suspicion as to the propriety of reducing the allowances. There is a question, however, which cannot be so easily disposed of. Why did the owners, in the face of last year's experience, not put a twelvemonth's provisions on board? Of course, nobody could have forced such additional provisions, and therefore the owners are not legally blameable; whether they are *morally* so is another question. But the fact is, Government should render such additional provisions absolutely binding on those who are connected with the Davis-Straits' fishery, now that the hazardous nature of the trade is so very obvious.

The case of the *Dee*, too, shows necessity of a very full supply of medicines suitable for the scurvy. Mr Littlejohn was perhaps well prepared; but there is reason to think that had his chest contained a larger quantity and a greater variety, the disease might have been mitigated.

The seamen who take out coffee and sugar at their own expense, will see, from this trying case, that they ought to be careful in laying in a large stock. Had those on board the *Dee* been amply provided with these articles, they would have felt the benefits of them when they had nothing to drink but water, half frozen, though newly melted from the snow.

The *Dee* arrived at Aberdeen on the fifth of May, which is thus noticed by the *Aberdeen Herald:*

The Dee arrived in the bay on the morning of the 5th, and at noon entered the harbour. The quay was crowded with anxious spectators; and, as the vessel neared the berth, the scene was truly heart-rending. The mourning relatives of the deceased seamen, though previously apprised of the unfortunate fate of those who were near and dear to them, seemed unwillingly to give credence to any testimony apart from a positive confirmation by those who had been eye-witnesses to their decease; or, believing the fact, seemed anxious to seize, with eager avidity, the earliest opportunity of taking a parting glance at the empty hammocks of the dead. Their weeping widows rushed on board with their helpless orphans in their arms, while parents and friends followed in equal grief. Of those who were privileged to meet their surviving relatives we need say nothing—their joy was great, but the detention of a few who were left at Stromness, led the expectant friends to give vent to the most frantic grief, and almost again to despair.

The total number of deaths on board the *Dee* was forty-six; of these, nine belonged to the *Thomas* of Dundee. Since the arrival of the *Dee* at Stromness, one of the invalids has died, thus making the loss no fewer than forty-seven souls. The number of widows and orphans left is very great, and their helpless condition renders them objects of great commiseration. It is to be hoped that a benevolent public will extend their liberality very freely in their behalf.

The 1830s EFFECTIVELY DEMONSTRATED the harsh vicissitudes of arctic whaling. The decade had begun with a season generally conceded to be the worst ever in the annals of British whaling. Of the ninety-one vessels that departed from English and Scottish ports for Davis Strait in the spring of 1830 nineteen were destroyed—more than one ship out of every five. The total catch of the fleet (excluding the wrecked whaleships) amounted to only 161 whales, an average of just over two per vessel. More than twenty ships returned "clean," having caught no whales at all.

In 1831, the year in which the *Hercules* of Aberdeen sailed to the Davis Strait fishery (chapter 3), ice conditions were less restrictive and less hazardous. Almost 400 whales were secured, but this was not enough to give whalemen and shipowners much encouragement.

Then came the extraordinary season of 1832, in which the combined efforts of the seventy-odd ships yielded 1,632 whales, the highest catch of any year, before or after, in the nineteenth century fishery, and an astonishing average of twenty-three whales per vessel. Another excellent catch was obtained in the next year, but a succession of severe ice years followed (chapters 5, 6). The total catch in 1836 was a mere sixty-six whales, the lowest catch recorded to that time, and a paltry four percent of the peak catch of 1832.

Such enormous fluctuations began to make shipowners and investors cautious. They had experienced heavy losses in capital equipment, losing almost forty ships during the decade, half of them in the first year or two, and the financial outlook was too uncertain to justify replacing these vessels or even keeping all of the remaining ones active in the whaling trade. Many owners began to employ their vessels in other activities, and some small ports drifted out of whaling altogether. Even the famous port of Hull, which had long dominated the arctic fishery, now relinquished its leadership to Peterhead; in 1840 it sent only two vessels whaling. By 1841 the British arctic whaling effort had declined to a fifth of what it had been ten years before. This

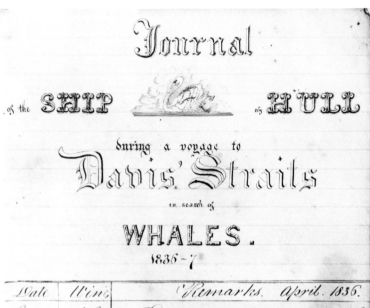

Journal

of the SHIP HULL

during a voyage to

Davis' Straits

in search of

WHALES.

1836-7

Date	Wind	Remarks. April. 1836.
Monday 11th	S.E.	Light winds & fine clear weather the whole of this day At 2 P.M. the Owners came on board and mustered the crew. The Ship lying in the Humber dock
Tuesday 12th	S.E.	Light winds and fine weather the crew employed in getting their clothes on board
Wednes. 13th	N. & W.	Fresh breezes & clear weather. At 4 P.M. the Pilot came on board warped ship out of the dock & ran down into Whitebooth roads & came to anchor in 7 fathom water
Thursday 14th	to N.W.	4 A.M. weighed anchor & ran out to sea at 10 A.M. the Pilot left the Ship

THE MISSING WHALERS.—The last of those unfortunate vessels arrived at Peterhead on the 29th ult. She is called the Swan of Hull, Captain Dring. Her crew has suffered dreadful hardships; a considerable number died.

RETURN OF THE LAST OF THE ICE-BOUND WHALERS.—The arrival of the Swan, the last of these unfortunate vessels, at Hull, on the 4th inst., is announced in another column. We revert to it now for the purpose of paying a well-merited compliment to Mr. Thomas Phillips, son of Mr. Henry Phillips, of the Old Brewery, Cardiff, for the professional skill and humanity evinced by him towards the crew of that vessel, which were happily the means under Providence of preserving the lives of so many as have survived through the privations and disease they suffered during their perilous voyage.

CREW OF THE SWAN.

On Sunday, July 2, 1837,

AT THREE O'CLOCK IN THE AFTERNOON,

A SERMON

WILL BE PREACHED

IN THE OPEN AIR,

Upon the Dock Green,
HULL

BY THE

Rev. R. Felvus,

AFTER WHICH

A Collection will be made for the benefit of the distressed Relatives of the Crew of the above-mentioned Fishing Ship.

Jabez Eden, Printer, White Horse Yard, Market-Place, Hull.

While the above Sermon was being Preached the unfortunate Ship "Swan" hove in sight.

Title page of a journal written on the Swan of Hull, one of the ships which survived the winter of 1836-37. (National Maritime Museum, London)

The Swan arrived at Hull during a sermon dedicated to her missing crew and distressed relatives. (National Maritime Museum, London) staggering reduction must have been accompanied by a number of economic adjustments in the shipping and manufacturing sectors.

Owing to the reduction of the Davis Strait whale population, which was the legacy of overhunting during the first third of the century, and to the plague of ruinous ice years, the 1830s marked an important turning point in British arctic whaling. Thereafter, whaling activity was markedly reduced; no subsequent year ever saw more than thirty British ships depart for Davis Strait, and the general trend during the next three quarters of a century was one of decline in number of voyages and in whales secured. The nature of the arctic whaling operation changed too, from this time onward. As yields dropped the tendency was to supplement whaling by sealing (first in the waters east of Greenland, and later in the vicinity of Newfoundland), to exploit other oil-bearing sea-mammals such as beluga whales, narwhals, and polar bears, and to seek animal products other than oil and baleen, including ivory, skins, and furs. Forthcoming decades would witness determined efforts to seek out the bowhead whale in untouched parts of the Eastern Canadian Arctic, to utilize Eskimo labour in the whaling, to winter on board ship in the region, to establish shore stations, and to develop more effective ships and equipment.

J. Henderson Lith. Ab. de.

The Aberdeen whaling master, William
Penny, who took Eenoolooapik to Scotland in
1839 and returned with him to Baffin Island
to initiate whaling in Cumberland Sound in
1840. (Scott Polar Research Insitute)

Eenoolooapik, the Eskimo who travelled to
Scotland on Penny's ship Neptune in 1839
and returned home the following spring on the
whaler Bon Accord. (Scott Polar Research
Institute)

Seven

A Young Eskimo of Considerable Intelligence

It was Nature's prompting that urged him to return to the land of his birth; for dreary and desolate though it might appear to others, its snow-clad hills and craggy cliffs were to him as the faces of familiar friends. ALEXANDER M'DONALD, 1841

THROUGH THE LONG HISTORY of European contact with the Atlantic shores of North America explorers have often seen fit to return to their home countries with tangible proof of discovery. What could be better than a live sample of the native population, in order to persuade government and merchant backers of the authenticity of the explorer's claims, and of the necessity of mounting a further expedition? No sooner had Columbus discovered the existence of the Bahama Islands and the West Indies than shiploads of Indians were being transported back to Spain as living evidence of the existence of a strange new land, and as slave labour. The heartless practice was continued throughout the Caribbean region and along the Atlantic seaboard of North America by explorers of several European nations, and the inhabitants of the arctic regions were not exempted. As early as 1502 Labrador Eskimos were carried to England by Sebastian Cabot, and in three voyages from 1576 to 1578 Martin Frobisher twice kidnapped Eskimos from Baffin Island and returned with them to England, where they fell ill and died. There was less of a need to send natives to Europe after the broad geographical features of the continent had become known and colonies established, but the traffic continued for one reason or another, sometimes to educate a few individuals, out of a humanitarian spirit, but more often to train them as interpreters, to be of practical assistance to traders and administrators in the New World. But they seldom fared well in the strange lands across the sea. In the eighteenth century at least seven Labrador Eskimos were taken to England, and all but one died of smallpox.

During the nineteenth century, whaling captains introduced a number of Eskimos to Britain and the United States. There is no evidence of force being used; these individuals appear to have chosen to make the voyages of their own volition, although doubtless it was the captain who usually took the initiative, offering free return passage in his ship, and promising to take the wary traveller under his protection and care while abroad. Sometimes these experiences were the natural outcome of bonds of friendship and trust that had developed between ship captain and native during several seasons of trade and co-operation in whaling, combined with a natural curiosity on the part of the Eskimo to know something about the place from which the white men came. In some cases, perhaps, the Eskimo adventurer

Lecture-Hall, Goodramgate, York.

THE TWO

ESQUIMAUX

OR YACKS,

Male and Female, brought home by Captain Parker, of the Ship *Truelove*, of Hull, from Nyatlick, in Cumberland Straits, on the West side of Davis' Straits,

WILL BE EXHIBITED

On Thursday and Friday, March 9th & 10th,

In the Lecture-Hall, York,

For Two Days only, previous to their return to their Native Country on the 20th Instant.

This interesting married couple, MEMIADLUK and UCKALUK, (whose respective ages are 17 and 15,) are the only inhabitants ever brought to England from the Western Coast. They have been visited by upwards of 12,000 persons in Hull, Manchester, Beverley, Driffield, &c.

THEY WILL APPEAR

IN THEIR NATIVE COSTUME,

With their Canoe, Hut, Bows and Arrows, &c.

From the Manchester Guardian, Jan. 5, 1848.

THE ESQUIMAUX.—Yesterday, we visited in the lecture-theatre of the Mechanic's Institution, one of those outlying varieties of the human family, not often seen in this country,—a young male and female Esquimaux, natives of Cumberland, on the south-west coast of Davis's Straits, in 65° 20 north latitude, and 67° west longitude. They were brought to this country by Captain Parker (of the whaling ship Truelove, of Hull), who, after having made upwards of twenty voyages to that coast, has had his sympathies so much awakened for a people perishing of hunger, that he has brought this couple hither, in order to bring the condition of the tribes throughout the west coast of Davis' Straits (which is British territory) under the notice of our people and government. It seems that while similar tribes of people along the whole of the east coast, or East Greenland, are living in comfort and plenty, under Danish rule, supplied by the Danes with implements of the chase and the fishery, and as happy as external circumstances can make them, the wretched people on the opposite coast of Baffin's Bay—speaking a dialect of the same language, and being to all appearance the same people—are in the most destitute condition; and that chiefly from want of fire-arms and other means of getting food by the chase on sea and land. Thousands of the wretched denizens of British territory on the west side of the bay, are now dependent on the charity of the captains and crews of whaling vessels, for the means of existence; and several of these captains, we learn, distribute amongst these poor polar savages, large quantities of food and clothing every voyage. Captain Penny of Aberdeen, nearly emptied his own clothes' chest to clothe them; and Captain Parker has expended upwards of £30 in procuring necessaries for them. The history of the two poor young creatures now here, is brief, but striking. Betrothed, as is the custom of the country, when children of four or five years of age, Memiadluk, the husband, is now only 17, and Uckaluk, the wife, only 15 years old. On the Truelove reaching the coast, Uckaluk, having just lost her mother, and being thus left an orphan, and without the means of subsistence (as all the possessions of the deceased are buried in the same grave) had nothing before her but to live as the dogs do, and perhaps to be devoured by them. Won by Captain Parker's long-tried kindness to the natives, she implored him to take her to England. He refused to take her alone, or to take a male and female, unless married. The two betrothed accordingly became man and wife the night before the ship sailed (the male being shorn of his long hair as a part of the ceremony), and were then received, together with Memiadluk's canoe, spear, small tent of skins, &c., on board the ship, and brought to this country. They were in the most filthy state, their skin coated with oil or grease, and covered with vermin, and, the girl especially much emaciated by want of food. They were cleansed, and new

sealskin garments given them, and they are now cleanly in habit, washing once or twice every day. They are, even to each other, taciturn; occasionally subject to great depression; but gentle, docile, grateful, and evidently much attached to Captain Parker. Both suffered considerably at first from sea-sickness, and having experienced every attention from Captain Parker, and Mr. Gedney, the ship's surgeon, Uckaluk went to the former with tears in her eyes, and said—"Uckaluk no father, no mother; Captain Parker be her father, doctor be her mother." In their own country these people eat all their food raw, and will devour from 2lb. to 8lb. or 9lb. of flesh daily; and during the voyage they ate quantities of raw leg of beef; but on reaching England they soon learned to eat cooked meat, and though taking 2lb. or 3lb. at first, their appetites are now not much greater than those of healthy labourers in this country. Being forewarned by Captain Parker, they never touch intoxicating liquors, and their only beverage is cold water. On recovering from sea-sickness, Memiadluk made himself useful on board, helping the sailors in various ways; and on reaching Hull, Uckaluk was taken to Captain Parker's house, and there soon learned to wash clothes, glass and crockery, clean knives and forks, &c. The exhibition is throughout a simple, but interesting one. Both male and female are clothed in a neat dress of sealskin; their stature is low; their colour dark, like that of the quadroon, but with long flowing black hair; their features seem a mixture of the Malay and the African, and in mild, sad expression, resemble the Hottentot. The male gets into his canoe, holds the paddle and poises the spear. A description, from which we have gleaned the above particulars, is given by Mr. Gedney, surgeon of the Truelove; and afterwards Captain Parker explains the condition of the people on both sides of Baffin's Bay, and draws that contrast between the Danish and the British rule, which is so little to the credit of our country. It appears that muskets and ammunition are all that are wanting to place these poor creatures in a condition of comparative comfort; and that the Danes, by doing this, by sending them medical men and missionaries, and building them wooden cabins, have not only increased their comforts, but succeeded in establishing a lucrative trade, taking from them whale and seal oil, and the skins of the seal, the bear, the fox, &c., and sending them guns and ammunition in return. Captain Parker is seeking by this exhibition to induce the British government to pursue some such kindly policy towards these territorial subjects of Queen Victoria; and as to the two Esquimaux, he is collecting for them with the proceeds of their exhibition, a good stock of these and other necessaries of Esquimaux life, preparatory to their return; and he assures us he shall take them back on his next whaling voyage, leaving this country about March next.

DAY EXHIBITION, from 2 to 4, ADMISSION, ONE SHILLING. EVENING EXHIBITION at Seven o'Clock, ADMISSION, SIXPENCE. SCHOOLS AND CHILDREN, HALF PRICE.

BROOKS AND LENG, PRINTER, BOWLALLEY-LANE, HULL.

counted on improving his stature among his own people by establishing a closer association with the whalemen and attaining a greater knowledge of their culture. And in many instances the whaling master considered that he would derive some benefits from the arrangement. Friendship, curiosity, and a degree of self-interest lay behind most of the Eskimo visits to Britain and the United States during the whaling period.

For those who chose to accompany a whaling captain to Europe or North America, a difficult but fascinating experience lay ahead. Like small farm children on their first visit to the big city, they would certainly be amazed and delighted at the wondrous things to behold, things quite beyond their most fanciful imaginings. But at the same time they would be confused and puzzled—perhaps even frightened—by the size and bustle of the cities, and the vast crowds of people. They would cautiously admire the technological marvels of the foreign countries—the horse-drawn trams, steam locomotives, large fleets of ships, tall buildings, and elevators—but the strangeness of the new surroundings and customs, and the difficulty of communicating, usually led to bewilderment and disorientation. Their mentors often treated them kindly, it is true, and people everywhere made a fuss over them. They were shown the sights in grand style; they toured the best shops; some were entertained by royalty and aristocracy. Many were invited to demonstrate their skill in paddling kayaks and throwing harpoons. Too often, though, they were displayed as sideshow freaks, and there were few among these tourists from the arctic who did not soon suffer from homesickness, alienation, and loneliness. An exhibition handbill of 1849 described a young man and woman brought from Cumberland Sound, Baffin Island in the previous season, as "taciturn; occasionally subject to great depression." The inference was that this was their natural condition. But it is much more likely that it was a product of their pitiful exploitation during a winter in England, during which (as the handbill boasted) they had "been visited by upwards of 12,000 persons in Hull, Manchester, Beverley, Driffield, &c."

Sadly but inevitably, a good many of the Eskimo visitors were

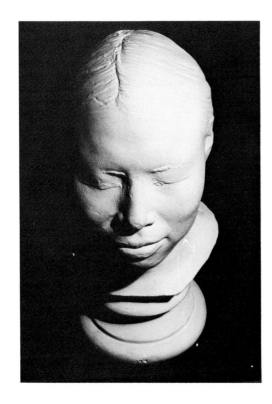

Far left: Handbill announcing an exhibition of two Eskimos who accompanied Captain Parker to England on the Whaler Truelove *in 1847. (Hull Museums)*

Top: Bust of Memiadluk, one of the Eskimos taken to England in 1847 by Captain Parker. (Hull Museums)

Right: Studio portrait of another Eskimo visitor to Great Britain. (Dundee Museums)

struck down by unfamiliar diseases and did not live to see their homeland again. For those who survived and returned, however, there were material riches to display, and fabulous experiences to relate time and time again to relatives and friends huddled in snowhouses and tents. Ironically, their descriptions of the wonders of an industrial society were so far beyond belief, and their stories so often repeated, that some of these returning travellers became known as great bores and incorrigible liars, after which they lost face and were shunned by their own people.

Eenoolooapik's British tour in 1839–40 was a result of his own initiative. For several years at Durban Harbour near Baffin Island's easternmost extremity at Davis Strait, the young Eskimo had attempted to join whaling vessels headed homeward to British ports but the entreaties of his mother had prevented him from leaving. Parental disapproval eventually waned, however, and when the Scottish whaling master William Penny invited Eenoo to accompany the *Neptune* to Aberdeen in 1839 he was at last allowed to depart on the magnificent adventure. For his part, Penny was anxious to obtain more precise information about the location of a deep arm of the sea to the southwest, called Tenudiackbeek, and said to abound in whales. It might be connected to, or synonymous with, the inlet penetrated for sixty leagues by John Davis two and a half centuries before but not subsequently visited by Europeans. Penny hoped that several months in Scotland would give Eenoo some fluency in English and enable him to direct the whaleship to the region in the following spring.

This published account of Eenoolooapik's experiences was written by Alexander M'Donald, who served as surgeon on board the *Bon Accord* in the following spring, when that vessel bore the young man back to Baffin Island. M'Donald later died on the Franklin Expedition of 1845.

October had now arrived, when all, save the rude denizens of the North, must leave those bleak, ice-bound shores; and Eenoo having obtained the consent of his friends, was taken on board the *Neptune*, accompanied by a number of his tribe. The Eskimo, with the exception of his mother, shewed little emotion at parting with him. With her, however, the case was far otherwise. Her first-born – now the chief guardian and support of her declining years – was about to visit a country and a clime far distant and unknown; to soujourn among a people whose language and manners he knew not; – and the promise of a stranger was her only guarantee for his safety. Under such circumstances it was not to be wondered at that maternal affection, implanted alike in the breast of the civilized and savage, should be displayed in all its power. Untrammelled by formal and frigid restraint, which oft-times checks the pure feelings of nature, and freezes the gushings-forth of the holiest affections, this unsophisticated Eskimo gave vent to her emotions in loud and prolonged bursts of wailing and tears. These expressions of her feelings lasted for some time, assuming various and somewhat extravagant phases, until, at last, in accordance with the peculiar manners of her country on such occasions, she laid bare her bosom, and invited him by an appeal, which, though silent, was irresistible, to kiss the warm breast which in infancy had suckled him; such being the last tender testimony of affection when the grave may prevent another meeting upon earth. At this touching scene Eenoo's resolution had well-nigh deserted him; but in a moment he rallied; the settled purpose of his soul was not now to be so easily subdued.

At this time Eenoolooapik was about twenty years of age, and might be considered, in his physical aspect, a fair specimen of the Eskimo race. But, as yet, his mental acquirements were of a very limited description. Doomed hitherto to pass his days amid those dismal solitudes of snow, where all his energies were requisite to provide for the wants of the passing hour, and where mental cultivation is unknown, it was scarcely to be expected that he would manifest much knowledge beyond what he had gathered in his wanderings, or what had been forced upon

This map, included in the published version of Eenoolooapik's adventures, shows both his place of birth (at Keimuksoke) and the settlement of Durban Harbour, from which he set out for England. Since it was drawn before Captain Penny had visited Cumberland Sound, it is far from accurate. (Metro Toronto Library)

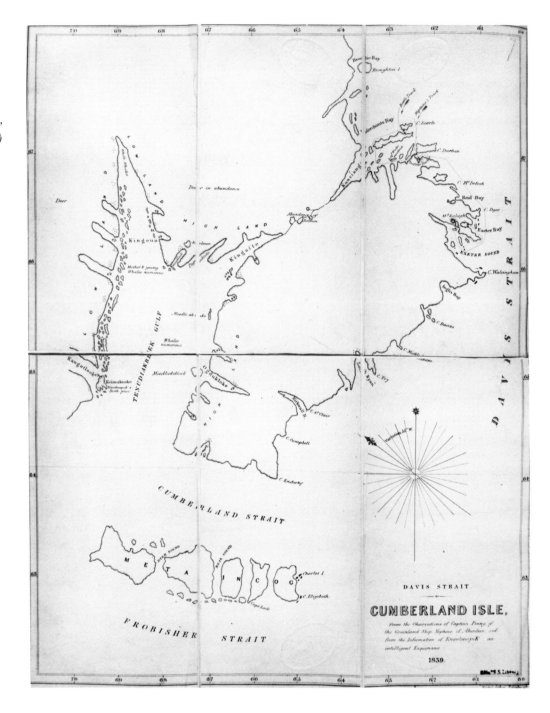

him by daily experience. And, indeed, if we except his geographical information, there was little to recommend him to the notice of our countrymen; but *that* being observed to be considerable, it was deemed of importance to have a better opportunity of learning the extent of the whale-fishery, but also be of value in a scientific point of view. ...

During the homeward passage every care was taken to instruct him in the usages of civilized society; and aided by the faculty of imitation, which he possessed in a very high degree, he adopted the manners of those around him with astonishing facility. Every attention was bestowed to prevent his morals being contaminated by intercourse with the vicious; and this was the more necessary, as the first impressions made upon a mind emerging from the gloom of savage ignorance, were likely to be permanent. His docility and the mildness of his disposition soon rendered him a general favourite; and the kindness which he in consequence experienced, no doubt contributed largely to the favourable opinion which he formed regarding the *Kudloonite* [white men]. At first he was rather averse to the change of dress which it was necessary he should adopt; for though it might please the eye and gratify his passion for embellishment, it was yet felt exceedingly inconvenient and irksome, and he would gladly have exchanged it for the loose furs to which he had been accustomed. He soon acquired habits of extreme personal cleanliness, – a circumstance the more surprising, that the Eskimo are generally very inattentive in that respect; but so complete was the revolution which his ideas underwent on this point of propriety, that in a short time he shewed an inclination to be rather fastidious than negligent.

An ample opportunity was now afforded Captain Penny for examining Eenoolooapik more minutely concerning Tenudiackbeek; and a kind of conventional language, composed of an intermixture of English and Eskimo, being established between them. Eenoo communicated many further particulars on that and other subjects; and

when language altogether failed him, he readily supplied its place by a rude drawing. In this manner, too, he represented his countrymen as engaged in encountering the various dangers of the chase, and thus conveyed to the mind a much more accurate idea than could possibly have been done by his imperfect verbal expression. ...

For the purpose of recording the duration of his absence from home, Eenoolooapik had recourse to the expedient of casting a knot upon a cord every morning when he arose; and when any unusual event happened, he cast a double knot to mark the period of its occurrence. On making the land, a double knot was cast; but his attention then became so much absorbed by the variety of extraordinary objects which were every hour presented to him, that the cord was laid aside and neglected. The frequent repetition of the words, Kudloonite! Kudloonite! (the white men! the white men!), pronounced in a slow drawling manner, which he always assumed when much pleased with any thing, was the only expression of admiration which escaped him. When proceeding close along the land, he remarked the distance of the houses from the water; a circumstance which surprised him so much, that he expressed great astonishment that people could live in such situations. He seemed, as yet, to have had no idea of the possibility of deriving the means of subsistence from any other source than the sea. Nor is it wonderful that he should have held this opinion, for his experience had been gathered from his own barren land, which produces little fitted for the purposes of man, save the moss for his winter's lamp.

He was first taken on shore on the coast of Caithness, at a place called the Castle of Mey. He expressed himself highly delighted with the sight of the Castle, and wished much to obtain a view of the interior of the building. This however, was denied him by the keeper of the mansion, who, with true Cerberus-like obstinacy, refused to allow the party even to walk round it.

A circumstance occurred here, which, while it afforded

considerable amusement to those about him, manifested the extreme simplicity of his ideas regarding the variety and extent of the animal creation. On seeing a cow and pony quietly grazing together, he stooped down and made towards them with the utmost caution, acting as he had been accustomed to do in the chase. When sufficiently near, he signified by a motion of his arm that they offered an excellent mark for an arrow. Observing the mirth of his companions, he returned and asked what kind of *deer* they were? or whether they were not *all the same as the Eskimo dogs?* Hitherto he had been ignorant of the existence of animals diverse from those with which he had been familiar; but being now undeceived, and at a loss for any other name by which to distinguish them, his mind reverted to those denizens of his own country as the only prototypes of the quadrupeds now before him. In the same way, when he first tasted a piece of codfish, he declared it was excellent *salmon*, or "all the same," as he expressed it; meaning, simply, that it was a fish bearing some resemblance to the salmon, with which he was well acquainted.

The *Neptune* had now arrived off Aberdeen, but the wind blowing from the eastward, a heavy sea ran upon the bar, and prevented her from taking the harbour. After contending for several days against an increasing gale, she was run up the Firth of Forth, and anchored under the island of Inchkeith. Here the *Sovereign*, steamship, of Aberdeen, was also lying, and Eenoo was taken on board that vessel. One of the passengers, not calculating upon Eenoo's keen sense of truth and right, and wishing to afford himself and others some amusement at the expense of the untutored Eskimo, took from his neck his gold watch-chain and threw it around that of Eenoo, who, somewhat surprised at the munificence of the stranger, asked if he meant to bestow it upon him. Being assured of this, he walked away, taking no further notice of the matter, till the gentleman becoming concerned for the safety of his property, began to insist for its return. To

this, however, Eenoo objected; saying, "you give me to take from me—not good—*Innuit* (the Eskimo) no do that:"—thus reading the gentleman a lecture in moral philosophy which he was not prepared to expect from such a quarter. Eenoo's firm refusal to deliver up the prize caused considerable merriment among the rest of the passengers, and he persisted in retaining it until the interposition of Captain Penny procured its immediate restoration.

This jest was no doubt attempted without the slightest intention of corrupting Eenoo, but it was obviously calculated to make a bad impression upon his inexperienced mind, as it tended to destroy those principles of rectitude which the Eskimo act upon among themselves. They are blamed, and not without reason, for being dishonest in their intercourse with us, and it is highly probable that this propensity was at first called into activity by trifling circumstances such as that now detailed. It requires little philosophy to account for this: for, finding deceit and falsehood practised towards themselves, and at the same time having strong temptations placed before them in the shape of articles useful to them, and unattainable from any other source, it is noways strange that they yield themselves up to the practice of secret cunning and appropriation.

Among the multiplicity of objects which Eenoolooapik saw in sailing along the coast, the lighthouses seemed to interest him the most; and of these the erection on the Bell-Rock attracted the greatest share of his attention. Concerning this structure, he asked whether it was one stone or rock? and whether it had not been brought from the land and placed where he saw it? He was easily informed of the mode of its erection and its use; and, indeed, the aptitude which he displayed in comprehending the nature of the many objects brought under his notice, was a matter of astonishment to all around him.

The weather having now moderated, they left the Firth and pursued their way back to Aberdeen, where they

arrived on the night of the eighth of November. Gaining the harbour while it was dark, there was no opportunity of witnessing how Eenoo would have expressed himself on a sudden view of the city bursting upon him....

There is a feeling of romantic interest associated in the minds of most people with the arrival of "strangers and foreigners" on our shores; and this principle of curiosity, as it is sometimes called, is heightened if the visitants be of a rude, uncivilized race. If it is not the same, it seems to be akin to the motive which induces us to visit a menagerie or a museum; although, when we gaze upon a fellow mortal in the uncouth aspect of barbarism, there may be more of *sympathy* mingled with the feeling than when we study the habits and instincts of the natural denizens of the forest. Mind is a subject of wonderful contemplation, whether exhibited in the refinement and science of civilized life, or in the wild, uncultivated manners of savage existence; and when a real "son of the desert" is brought amongst us, we naturally feel a strong desire to witness the workings of his untutored reason, and the development and display of energies which have slumbered till the moment he is ushered into the midst of civilization.

Eenoolooapik was greatly impressed by the lighthouses on remote parts of the Scottish coast and eager to know how they were built. (Metro Toronto Library)

A month after being "among the fur-clad savages" of Baffin Island, Eenoo was "in the midst of a civilized and refined community," nineteenth century Aberdeen. (Metro Toronto Library)

The news of the arrival of an Eskimo in Aberdeen produced considerable sensation among all classes of the inhabitants; and on the following day great numbers of people collected on the way for the purpose of obtaining a sight of Eenoolooapik. The cabin of the *Neptune*, too, was crowded with visitors, and Eenoo was thus subjected to much that was disagreeable and foreign to his constitution, in the confinement and increasing heat of the narrow accommodation. He was, in consequence of being thus exposed to an overheated and vitiated atmosphere, seized with a pulmonary affection, which, though slight at first, the humidity and somewhat variable nature of our climate tended to aggravate....

He was now transferred from the *Neptune* to the more comfortable accommodation of a town residence; and the same facility of comprehension was displayed by him in reference to every thing to which he was introduced. Shortly after his arrival he was invited to a dinner party, given expressly for the purpose of ascertaining how he would conduct himself amongst the higher and more fashionable circles of society, before an opportunity had been afforded him of becoming acquainted with the forms which are there observed. On this occasion every thing was exhibited which was likely to astonish him and elicit the latent feelings of delight, which must, unquestionably, have possessed his soul. So far from being in the slightest degree confused, he acquitted himself in a manner which surprised every one present. The faculty of imitation, which, as we have before noticed, he possessed in a high state of development, enabled him to copy the manners of those around him with such promptitude and precision, that it would have been difficult for one unacquainted with the fact to have told that he had been accustomed to move in a different sphere of life. The smile, the bow, and even the slightest gesture, he imitated with the most minute correctness. He expressed no astonishment at anything which occurred, until the table was exposed on the removal of the cloth; struck by its extent and beauty, he uttered an exclamation of surprise, and set about examining its structure and qualities.

That the propriety of Eenoo's behaviour on this occasion depended principally on his power of imitation, may be proved from the following circumstance: With the view of ascertaining how far his conduct might be attributed to this faculty, one of the gentlemen at the party purposely committed a breach of etiquette, and was immediately followed to the very letter, in his unusual course, by Eenoolooapik. But, being made aware of his error, and of

the imposition which was practising upon him, without allowing his self-possession to be at all disturbed, he looked around, and after consulting the countenances of the various individuals, he readily concluded who he ought to imitate.

He was next taken out for a short distance to the country. He expressed himself as gratified with the appearance which it presented, and contrasted it with the aspect of his own sterile land. The trees, especially, astonished him by their magnitude; and he amused himself in measuring the circumference of several of them, and in comparing them with the stunted shrubs of the *West Land* – as he had been taught by the sailors to denominate the country of his birth. He displayed considerable anxiety to be informed concerning the nature of every strange object; and, in return, he was very ready to communicate such knowledge as he possessed, in regard to the productions of his native clime, whenever an opportunity occurred for his doing so. It may be here remarked, however, that although he seemed interested about every thing which he saw, he maintained the utmost coolness and deliberation in examining whatever attracted his notice. The same perfect composure and gravity marked his intercourse with the various individuals whom he met; and, as yet, he was equally at home with every person, knowing none of the ordinary distinctions of society.

The change of circumstances which Eenoolooapik had undergone, was perhaps as great and rapid as can well be conceived. A month ago, and he had been among the fur-clad savages of Durban, a member of their tribe, and a follower of their customs; and now, he was an object of attraction and interest in the midst of a civilized and refined community. It is difficult to imagine the process of thought which must have passed through his mind within this brief period; and his ready intelligence and perfect equanimity are still more curious and interesting phenomena. It might have been expected that one whose life had hitherto been spent amidst the bleak scenery of an arctic shore, where little save the bare rock, the withered lichen, or the eternal snow, meets the eye at every turn, would have been altogether bewildered by such a transition as that which Eenoolooapik had just experienced.... The isles of the Pacific have sent of their sons to see the fatherland of the faithful missionary,–the dark children of Africa have come to behold and bless the birth-place of liberty to the captive negro,–the simple Hindoo, and the stern Indian, may have trod our soil and wondered at our science,–but all these have the remembrance of much that was lovely and luxuriant in their own fair and fertile homes. Eenoolooapik's memory had no such beautiful resting-places on which to repose and expand itself.... The towering cliffs of the stormy north may display much of grandeur and magnificence, but the cheerless snow-hut and the icy ocean can call forth few associations of repose, and could have done little to prepare Eenoolooapik's mind for the refinement into which he had been ushered.

A few days after the arrival of the *Neptune*, Captain Penny, at the urgent solicitations of his numerous friends, allowed Eenoolooapik to display his dexterity in the management of his canoe on the river Dee. On this occasion he was with the greatest difficulty prevailed upon to exhibit himself in his native costume; but so changed were his opinions on the subject of dress, that he only did so, on being assured that he would never be asked to put it on again. The day happened to be exceedingly warm for the season of the year; and Eenoo, ambitious of shewing his expertness, exerted himself to the utmost of his power. He became considerably overheated in consequence of the severe exercise and the warm nature of his dress, and the pulmonary affection from which, as before mentioned, he was suffering, was thereby aggravated. Its alleviation, too, was afterwards prevented by the imprudent manner in which he exposed himself in the open air; for he resisted every entreaty to remain in the house, unless when Captain Penny was present with him. His disease, however, at length assumed such an alarming form, as not only to render confinement to his apartment imperative, but even to threaten his existence.

Arrangements had been made for instructing him in such elementary branches of education as it seemed he was most likely to acquire with ease, and also for teaching him the art of boat-building, which it was thought would be highly useful to him. But the excellent and praiseworthy intentions of his friends were unfortunately frustrated by the very serious aspect which his malady had now assumed.

The disease from which he was suffering was an inflammatory affection of the lungs. It was extremely severe, but it presented no other remarkable peculiarity. The Eskimo, even in their own country, are very liable to such affections during the summer months. They do not use any remedial measures, at least any which can properly be called so; but nature generally performs a cure by means of copious bleedings at the nose. They place implicit reliance on the powers of the *angkuts* [shamans], who, when they visit a patient, bind up his eyes and utter some mystical sounds, by which, it is believed, they invoke the Great Spirit on behalf of the sufferer. The patient gets better in the way already noticed, and the *angkut* receives the credit of the cure, and some substantial present for his services. Reared in the belief of the efficacy of their incantations, Eenoolooapik strongly objected to medical treatment; nor would he at all submit himself to it, until assured that it was the only means of saving his life. This being strongly represented to him, he was at length persuaded to allow himself to be bled; but it was evident that he suffered the bleeding rather from the remonstrances of those about him, than from any idea that his disease would be subdued by such a remedy.

By the means adopted by his medical attendant, he

recovered so far, that in about fourteen days he was able to leave his bed. He was now satisfied that the treatment which he had undergone had been of some efficacy; but his faith in the *angkuts* was not on that account in the least shaken. When speaking of their pretensions, he related that on some previous occasion he had been very ill, and an *angkut* had been called to see him, who, as he said, attempted to cure him "by much talking"; but he stated it as an *extraordinary circumstance* that *his* disease had not been quelled by the power of the incantation. He did not perceive that this case militated against his own belief, for he evidently wished to convey a very favourable idea regarding the success of the *angkut* practice.

As soon as Eenoolooapik's health was sufficiently improved, he was taken out to spend a day at the house of a gentleman who had shown a great deal of interest in him. Here he rendered himself amusing by the representations which he gave of the winter employments of the Eskimo. His ignorance of our language prevented him from using it in expressing his meaning; but his pantomimic representation of the seal-hunting and similar pursuits, is said to have conveyed to the minds of the beholders a very lively conception of what he meant to describe. The keenness with which he entered upon this exhibition, produced so much excitement in his yet weak frame, that a relapse of his complaint immediately followed; and it was attended with even more alarming symptoms than had been manifested in his first attack.

He was already so much weakened by the depletion which had been necessary to subdue his first illness, that he was little able to bear the remedies which were requisite in this new attack. For nearly three long and weary months he was confined to his bed,—his frame shattered, his strength wasted, and his mental energies impaired. He was, in short, brought to the very brink of the grave. After the excitement which attends the earlier stages of his affection had passed away, he sunk till his life trembled in the balance. He lay motionless and apparently unconscious of what was going on around him. His extremities were cold, his eyes sunken, and the expression of his countenance ghastly—the powers of nature seemed exhausted: yet he rallied; but his convalescence was protracted, and the slightest exposure tended to produce a relapse.

For a long time he suffered with the most exemplary patience, and not even a murmur escaped him; but at length he became enervated by continued pain, and the thought that he was dying began to steal over him. He then displayed great anxiety on account of his mother, and was distressed by the reflection that he would never see her again. On one occasion this feeling operated so powerfully upon him, that he cried bitterly, and reflected upon Captain Penny for bringing him away.

This solicitude on account of his mother was nothing more than we should have expected, considering the duty which is required from an Eskimo to his maternal parent. Her support in old age entirely devolves upon him, and when deprived of this, her condition is miserable in the extreme. Hence, to be childless is among them considered one of the greatest misfortunes imaginable; and when such happens, it is common for them to adopt the children of others, in order that they may not be left destitute in the evening of their days, when they have become unfit for the active duties of life. The practice in question prevails most extensively in regard to boys, they being most useful; but instances are not unknown where girls have been adopted; and even the exchange of children is not uncommon, apparently for the purpose of preserving an equal balance of the sexes in a family.

During Eenoo's recovery it was found necessary to aid the prostrate powers of his constitution by means of wine and other stimulants. When this practice was first adopted he seemed rather averse to take them, but by and by his disposition towards them assumed a more friendly aspect. Having, apparently, discovered their cheering influence, he was in the habit of slipping out of bed, when unperceived, and tasting a little of the inspiring liquid. It was once or twice remarked that he shewed an unusual flow of spirits, but his previous habits having been exceedingly temperate, the cause of his exhilaration was never suspected. He was one day, however, caught in the act, which put an end to all doubt on the subject, and also to his private indulgence.

There was perhaps nothing uncommon in this temporary wine-bibbing propensity, but it will be allowed that in the following respect his taste was somewhat peculiar. It had been found necessary to administer castor oil to him on various occasions, and instead of loathing this nauseous draught, as is usual with patients, in this country at least, he was always willing, and even rather anxious, for its copious administration.

He was very observant of the means taken for the cure of his disease, and particularly anxious to learn the nature of the indications which the pulse afforded. He would narrowly watch till the physician withdrew his finger, and then he would pounce upon the pulse, as if fearful of losing the opportunity of examining its mysterious indications.

Returning health brought back with it his old feelings and associations, and hope with its ever-cheering beam again illumined his soul, and caused him to forget his recent sufferings. Indulging in the anticipations of future pastime amid the wild crags of his wintry home, he one day asked what time he was to get a gun which Captain Penny had promised him. He was yet scarcely able to leave his couch, and Captain Penny, thinking that it might relieve the tedium of his weary hours, immediately procured the fowling-piece for him. When presented with it he expressed himself highly delighted, and after having examined it sufficiently it was set aside. The following morning he was observed to be exceedingly languid, and apparently much worse. On inquiry being made as to how

he had passed the night, he confessed that he had crawled from his bed, and spent several hours in examining his gun; which he had been enabled to do at the window, in consequence of its being moonlight at the time. The bad effects of this imprudent exposure, however, soon wore off, and his recovery went on progressively.

Eenoolooapik's illness was a source of the utmost anxiety to Captain Penny, who had, without any reservation, engaged to restore him to his friends in safety. Indeed he had even gone so far as to promise, that if any evil befell Eenoo, he would deliver himself up to be dealt with as they should think fit. This engagement, although made in the best possible spirit, was yet of very questionable propriety; for, had Eenoo died, quiet and inoffensive as the Eskimo generally are, we are by no means satisfied of Captain Penny's safety, if ever he should have come within the range of their power. But, altogether apart from his solicitude on that account, the attention and kindness which he shewed to Eenoo during his illness, were of a character which commands our admiration, and does him the highest credit. He watched and tended him while he lay sick and powerless; he relieved the tedium and monotony of his couch in convalescence; and, in short, he was, as Eenoo himself expressed it, "all the same to him as a mother."...

...On his health being now in some degree re-established, the process of teaching him to read was commenced. He mastered the alphabet with great readiness, but here his literary attainments terminated. He had evidently no relish for such pursuits, for he could not perceive any advantage which would afterwards accrue to him from the knowledge of letters. It was chiefly by this *prospective* principle that he was guided in every thing which he set about learning or acquiring. If he did not see that the subject of study or acquisition would be of future utility, he could not be persuaded to bestow attention upon it. When any *toy* accidentally came into his possession, he would examine it with great curiosity and care, but after discovering that none of the practical purposes of life, so far as known to him, could be served by it, it was soon thrown aside as useless. On the other hand, if he got any thing which he judged might afterwards be turned to account in his own simple avocations at home, he hoarded it up with the greatest eagerness.

With this indifference to literary study may, however, be contrasted his partiality to drawing. This peculiarity of his disposition must have resulted from the predominance, in him, of those faculties of the mind on which that art depends; for we can scarcely conceive any thing which was likely to be of less service to him amid the desolate scenes of his arctic home. He had from the first shewn an aptitude in that art which, if he had remained in this country and received instruction, might soon have rendered him an adept in the use of the pencil.

The restoration of his health being now so far perfected as to allow him to be taken out, he was again introduced to every thing which was likely to interest or inform him, and he displayed the same readiness of comprehension which had previously characterised him in the examination of every thing which was new and wonderful. When his attention was directed to any strange object or piece of mechanism, if he had ever seen any thing bearing even the most distant resemblance to it, he would instantly refer to that, and state that what he now saw was *"siniagout,"* or "almost the same." Frequently when any thing curious was shewn him, he would examine it without making any remark, so that it appeared at the time to have made little impression upon him, but he would afterwards revert to the subject, and ask innumerable questions concerning it. On one occasion he was taken to a manufactory and first shewn the cotton in its raw state, and then he was conducted through the various apartments of the establishment and shewn the changes which the machinery

effected on the substance which he had first seen. When he found it brought into the condition of fine thread, he took hold of the wristband of his shirt, and asked if that was not "all the same?" He was several times taken to the theatre and other places of public amusement, on which occasions he seemed to enter with all his soul into the nature of the scene. The theatre, in particular, was a source of much gratification to him, for it seemed as if he fully comprehended the exhibitions, and could judge of the assumed character and language of the various performers. He stated that the Eskimo have similar pastimes, but of course on a more diminutive and less refined scale....

The latter part of Eenoolooapik's residence in Aberdeen was not characterised by any occurrence of moment.... As his acquaintance with the world increased, he became more retiring and bashful in the presence of those with whom he was not intimate; whereas, at first, he was equally pleased with and communicative to all. This change, however, did not go so far as to affect the propriety of his behaviour, which, from being simply imitation of the actions of others, had now become a habit with him.

Ever since his transference to the *Neptune*, he had been accustomed to the use of our food, which did not seem to be productive of any injurious consequences to him, although he had never even tasted any thing of a vegetable nature before. It would appear, however, from some circumstances observed during his illness that animal food was best suited to his constitution, as it was given to him with decided benefit when he was in a condition very different from that usually requiring its administration. He was, at first, in the habit of taking his food in a half raw state, but in a short time his taste in this respect underwent a complete change, and he refused it when presented to him in that condition, declaring that it had got *"oko too little,"* or "too little heat." He shewed no disposition to engorge himself in the manner so common among the Eskimo but on the contrary he was exceedingly abstemious and moderate. Indeed, he shewed none of those fierce and ungovernable passions which characterise man in his savage condition, but, on the contrary, he was mild and gentle in his nature, and modest, and even delicate, in his intercourse with female society. The attention which he received from the inhabitants of Aberdeen, and in particular from those of them connected with the *Neptune*, both on his arrival and during his residence among them, was not more flattering to the dependent stranger, than honourable and praiseworthy to the parties bestowing it. Of their attention and kindness he seemed duly sensible, at least he rewarded them with much deference and respect, and general amiability of character. He once or twice displayed some little bursts of self-will, which owed their origin to over-indulgence during his illness; but these were of short duration, and soon gave place to his usual blandness and equanimity.

He had learned to abstain from his usual amusements on the Sabbath, and, previous to his leaving, he was taken to church, where he conducted himself with the utmost propriety, and followed the external ceremonies of the worship as if he had been accustomed to them all his lifetime. His instruction in religion had also been prevented by his long and dangerous illness, and it is questionable whether he had any understanding of the forms and observances in which he so readily joined. Nor was there now an opportunity left for his information, as his mind had become so much absorbed by preparation for his return, that nothing was heeded by him which had not a reference to his approaching departure.

In consequency of the information which he had given, it was determined that the *Bon Accord*, whaleship (to which vessel Captain Penny had in the meantime been appointed), should, in the course of her voyage, examine Tenudiackbeek, and test the truth of Eenoolooapik's statements.... Having agreed to accompany Captain Penny in the expedition, the writer now became personally acquainted with Eenoolooapik.

As the time of Eenoolooapik's departure drew near, he assumed a more business-like air, and employed himself in collecting such things as he thought would convey to the minds of his countrymen some idea of the wonders he had witnessed; and also in providing himself with numerous articles which would be useful to him in his various pursuits at home, and much more efficient than any thing which the rude arts and limited resources of the Eskimo could supply.... The Lords of the Treasury placed twenty pounds at the disposal of Eenoo's friends in Aberdeen, for the purpose of assisting in procuring whatever might be considered necessary to establish him in his native country in more comfortable circumstances than he had formerly enjoyed. And well provided indeed he became, for no cost was spared by his friends in furnishing him with every thing that was useful or desired by him. Fowling-pieces, with powder and shot, edge-tools, of various kinds, culinary utensils, and clothing in abundance, formed part of his miscellaneous acquisitions.

Although he had experienced so much favour and kindness, he did not seem to feel any reluctance to leave this country, but rather looked forward to his departure with pleasure. But, indeed, this need not be wondered at when we think of the sickness and sufferings he had endured. He was fully sensible that the climate did not agree with him, as he was constantly annoyed with cough, and, in fact, his health was never so thoroughly confirmed as to shew that he was becoming at all acclimated. It is highly probable that he would not have survived long here, but would have fallen a sacrifice to the insalubrious influences of our moist, inconstant atmosphere, and have found a grave where he had come to worship at the temple of knowledge.

T HE LONG WINTER among unfamiliar surroundings and strange people came at last to an end, and on the first day of April 1840 Eenoolooapik left Aberdeen for his homeland on the *Bon Accord*, under William Penny. When the ship came up with icebergs a month later he began to regain the natural ebullience and vigour that had flowed away during dreary days in Scotland. He responded with unusual enthusiasm to instruction in writing and drawing, and took great interest in the language and implements of Greenland Eskimos encountered at Rifkol, Disko, and Upernavik. The ship continued northward and on 23 June narrowly escaped destruction during a gale that drove the pack in upon the whaling fleet and swiftly crushed the nearby whaler *Hecla* against the floe edge. If Eenoo had come to suspect that white people sometimes behave in curious and even irrational ways, his suspicions must have been strengthened considerably on the following day, after the storm had subsided, when about 150 men converged on the wreck of the *Hecla*, extracted a large cask of rum from the shattered hulk, and indulged in a "Baffin Fair" on the ice. Pressing every available container into service, including hats and sea boots, they went at the rum with a determination born of long abstinence and heightened by relief at having come safely through a perilous situation. The thirsty throng consumed the rum rapidly and in excessive quantities, with the predictable and sought-after result of widespread and profound intoxication. Surgeon M'Donald employed buckets of icy water to assist seamen to regain some portion of their senses, while officers on an adjacent whaler took firmer measures, hoisting inebriates from the yardarm by block and tackle and dropping them repeatedly into the icy water alongside until a modicum of sobriety returned. The spectacle of some men hauling others into the air with ropes and plunging them into the sea must have seemed astonishing to Eenoolooapik,

even after a winter's residence among the Scots. Surely it was a sport which would rival caber-tossing in its implausibility. Surgeon M'Donald, with a detachment born of familiarity and a degree of understatement appropriate to his position, merely recorded, "such is the infatuation of many of our sailors, that whenever an opportunity occurs for indulging in intoxicating liquors, they will embrace it under whatever circumstances."

Ice prevented the fleet from crossing Melville Bay in July and one by one the ships turned south again to look for a passage to the West Land at lower latitude. The *Bon Accord*, one of the last to leave, retraced her path about 700 miles before managing to find an opening in the pack at 65°N on 27 July and, with the help of Eenoolooapik, Penny took the vessel at last into Tenudiackbeek, the region of Eenoo's birth.

Hogarth's Sound was the name Penny gave to the large indentation; another whaling master called it Northumberland Inlet. It later became known variously as Cumberland Strait, Cumberland Inlet, or Cumberland Gulf, and is now officially Cumberland Sound. Since its discovery by John Davis in 1585 this body of water had remained beyond the sphere of European maritime activity, lying undisturbed behind a fringing barrier of pack ice, which discouraged vessels from approaching. Now its isolation came quickly to an end. Penny's rediscovery of the place was no secret; several other captains, familiar with the descriptions made public during Eenoo's sojourn in Scotland, had remained close to the *Bon Accord* as she sought the virgin whaling ground of Tenudiackbeek. Within a few years Cumberland Sound would become a principal focus of West Side whaling.

As Captain Penny explored the unfamiliar waters during the month of August, Eenoolooapik met Eskimo friends and relatives at a number of localities. With sophistication and restraint he showed some of his new possessions, distributed gifts, demonstrated the effectiveness of his fowling piece, and told of his experiences in Greenland and Britain. At the settlement of Noodlook he lost no time in charming a young lady and soon requested the ship's surgeon to officiate at a

Christian marriage ceremony—which, however, the surgeon was not empowered to do. At last the time came to take his departure. The ship's officers loaded a skiff with his "immense quantity of indescribable articles," deposited him ashore among his relatives, and bade him farewell.

Whether his material wealth and broad experience worked to his advantage among his countrymen, we cannot know. In some other introductions of Eskimos to Britain and the United States, however, the returning visitors found themselves uncomfortably situated between two cultures, unwilling to endure again the demanding conditions of life on the land and unable to satisfy the aspirations and tastes cultivated in a more complex society, a dilemma that has often confronted the Eskimo people since that time.

Eenoolooapik found himself a wife and sired a child, but died of consumption in 1847.

Flensers at work on a whale with mollie boats alongside. The baleen in the whale's mouth can be clearly seen. (Dundee Museums)

Hot and Greasy Blood

*I think there are few men who having once seen the exciting scene of
a whale hunt, would for an instant prefer their beds to the pleasure of seeing
it again.* ROBERT GOODSIR, 1850

FEW OF THE ARCTIC WHALING MASTERS who pursued their dangerous business among the ice floes of Baffin Bay year after year had the time or inclination to set down their experiences in writing. Their exploits were eagerly reported in the newspapers of the whaling ports, after dockside interviews when the ships arrived home, but they were given less attention in other cities and regions, and consequently the achievements of the whalemen were not as well known among the public at large as they perhaps deserved to be. A great body of firsthand knowledge of the arctic regions lay in the minds, the logbooks, and on the chart tables of the veteran whalemen, but this information served the whaling fleet alone and was not available to other sectors of society. Nor was it expected to be. The tasks of charting coastlines, examining natural history, cataloguing observations, and publishing scientific results, were considered to be the responsibility of discovery expeditions sponsored by government, private benefactors, or merchant backers. So during the nineteenth century the whaling grounds of the Davis Strait region were traversed not only by the whaling vessels of several nations, but also by exploration ships seeking to discover new lands, resources, and routeways, to attain specific geographical objectives, and to gain scientific knowledge of northern environments. Whaling activity and exploratory enterprise proceeded independently of one another, with little co-operation or exchange of information.

The principal objective of arctic explorers since the late sixteenth century had been to find a sea route from the Atlantic to the Pacific that passed north of the North American mainland—a so-called Northwest Passage. Like adding pieces to a jigsaw puzzle, a succession of maritime probes in Baffin Bay, Hudson Bay, and Foxe Basin, several overland expeditions funnelling through the subarctic regions, and a number of coastal boat voyages along the northern mainland coast, had by 1840 given an indication of the composite picture, although many details remained to be filled in.

Maps showed two parallel, overlapping corridors of known coastline trending east-west between Baffin Bay and Bering Strait, separate and unconnected. The northernmost one, leading westward from Lancaster Sound through Barrow Strait and Viscount Melville Sound, appeared to be a Northwest Passage in itself, but it ended in the old, impenetrable pack ice of the Arctic Ocean and had therefore proved to be a cul-de sac. Edward Parry and others had fixed the position of many headlands and coastal segments along the margins of this routeway, and their work suggested the presence of islands separated by straits leading both north and south. The southern corridor snaked along the mainland coast and appeared to offer a continuous passage from longitude 95°w (a little west of Boothia Peninsula) to Bering Strait, the doorway into the Pacific.

The geographical problem was simply to find a navigable connection between the Lancaster Sound and mainland-coast corridors, and it was widely believed that one well-equipped expedition could accomplish this and make the first voyage through the waterway. The British Admiralty rose to the challenge, chose Sir John Franklin as commander, and despatched HMS *Erebus* and HMS *Terror* from England in 1845. In late July, when the expedition vessels communicated with the whalers *Enterprise* and *Prince of Wales* at approximately 75°N, 66°w in Melville Bay, their crews were in good health, their spirits high, and their officers confident of success in the task ahead. This was the last time any of Franklin's men were seen alive by Europeans.

As years passed with no further news of the expedition Britain's expectations of quick success faded, to be replaced by growing concern about the safety of the 129 missing men. A few cursory attempts were made by whaling captains to look for signs of the expedition in Lancaster and Jones sounds in 1847 and 1848, and in the latter year the Admiralty sent out the first official searching expeditions. In a gigantic pincer movement reaching halfway round the world Thomas Moore sailed HMS *Plover* on a nine-month voyage to Bering Strait, the western end of the yet-to-be-traversed Northwest Passage, while James Ross and Edward Bird took HMS *Enterprise* and HMS *Investigator* along Franklin's original path into Lancaster Sound at the eastern end of the passage. Both expeditions were prepared to spend the winter in the arctic regions.

It must have seemed likely that such energetic measures would result in the finding of the *Erebus* and *Terror* and the

The arctic explorer Sir John Franklin, whose expedition to find the Northwest Passage disappeared in the summer of 1845. (Public Archives Canada, C-1352)

saving of Franklin's men, but they did not. These and dozens of other search expeditions during the next decade were to be defeated, as Franklin and his crews had been, by the vastness and complexity of the arctic archipelago, the severity of the region's climate and ice conditions, and the difficulties of

FRANKLIN'S LAST VOYAGE

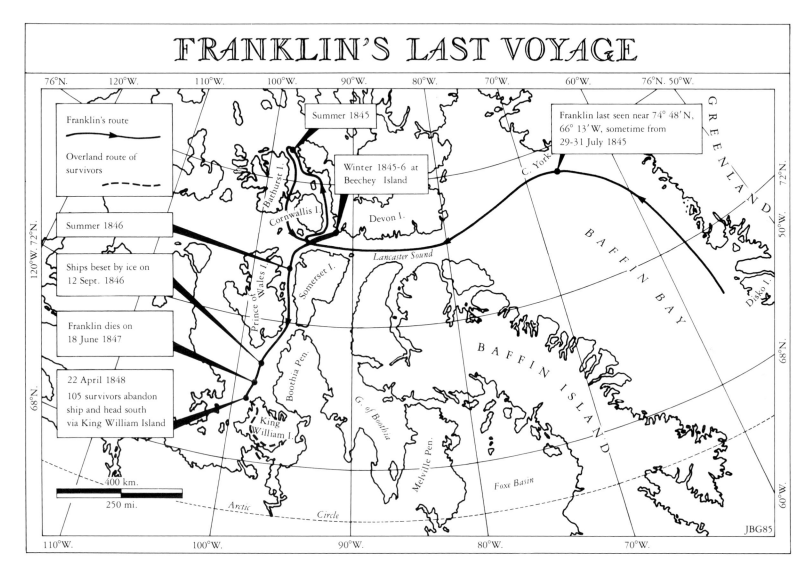

Franklin's route

Overland route of survivors

Summer 1846

Ships beset by ice on 12 Sept. 1846

Franklin dies on 18 June 1847

22 April 1848
105 survivors abandon ship and head south via King William Island

Summer 1845

Winter 1845-6 at Beechey Island

Franklin last seen near 74° 48′ N, 66° 13′ W, sometime from 29-31 July 1845

GREENLAND

BAFFIN BAY

BAFFIN ISLAND

Bathurst I.

Cornwallis I.

Devon I.

Lancaster Sound

Prince of Wales I.

Somerset I.

Boothia Pen.

G. of Boothia

Melville Pen.

King William I.

Foxe Basin

Disko I.

C. York

400 km.

250 mi.

Arctic Circle

JBG85

preventing scurvy. In addition, the lack of an effective communication system to co-ordinate the operations of ships, boats, and sled parties greatly hindered the search efforts.

In general whalemen were not closely involved in the search for Franklin. As their wages depended partly on the amount of whale oil and baleen they could secure in the short arctic summer they were reluctant to depart from their usual

itineraries unless compensation was paid. Nevertheless one captain did contribute significantly to the search. William Penny, the same man who had taken Eenoolooapik to Scotland in 1839 and initiated whaling in Cumberland Sound in 1840, had made the first efforts to find traces of the missing ships in 1847, but had been blocked at the mouth of Lancaster Sound by fast ice. Now in 1849, as he set out on the *Advice* of Dundee on another whaling voyage, he intended to try again. On board was Robert Goodsir, who was eager to learn something about the fate of his brother Harry, assistant-surgeon on HMS *Erebus*.

Goodsir's published narrative of the voyage is a fascinating record of whaling operations. Like surgeon John Wanless, aboard the *Thomas* in 1834 (chapter 4), Goodsir was not content to remain a passive observer. He made strong efforts to participate in the various activities, and recorded them in careful detail. Frustrated at the outset by being marooned on the deck when the whaleboats were lowered to pursue whales, he later managed to join a boat crew and was then able to experience the same risks and fears as the whalemen.

The *Advice* cleared Stromness on 17 March, rounded Cape Farewell a month later, and pressed northward into Davis Strait. Progress along the Greenland coast was punctuated as usual by brief visits to Danish settlements. Like most of the sailors, Goodsir had had the foresight to purchase a modest supply of trade goods before leaving Scotland, in order to "troak," or trade, with the Greenlanders for the handicrafts they customarily offered on board the whalers — "sealskins, sealskin trousers, caps, slippers, gloves, and tobacco-bags or 'doises.'" When kayaks came off to the ship at the Whalefish Islands in Disko Bay the surgeon confidently produced his carefully chosen, store-bought wares, including clasp knives and large sailmakers' needles, fully expecting to command the market with these valuable commodities. To his amazement his potential consumers turned up their noses in contempt, rejected his large needles, and moved quickly to other whalemen to barter eagerly for small sewing needles and cotton handkerchiefs in the gaudiest colours. Goodsir's market research had been inadequate. After almost two centuries of trade contact with European vessels the Greenlanders had come to know perfectly well what they could obtain, what they wanted most, and what they did not want at all. As for his sailmaker's needles, the natives "would scarcely have them as a present."

Goodsir noticed that the leather for the slippers, far from being a native product, was sent over from Denmark for the Greenlanders to cut, sew, embroider, and trim with fur. This appears to have been a rare reversal of accepted mercantilist procedures in which raw materials were normally exported from a colony to the mother country for processing, manufacture, and sale.

Continuing north, the *Advice* took her first whales — four of them — off Hare Island and Black Hook. These were merely preliminary benefits, but an intensity of purpose now pervaded the shipboard routine and everything became secondary to the attainment of the more productive West Side whaling grounds and the securing of oil and baleen. During June the ship worked through the pack ice of Melville Bay, emerging on 4 July in the North Water beyond Cape York. The West Land was sighted four days later.

We had a distant sight of the west coast of Baffin Bay, about lat. 76°N, on the eighth of July, being a part of north Devon Island. We ran past the mouth of Lancaster Sound with a strong breeze, and occasional heavy squalls. The ice we passed during the day was much heavier than any we had seen on the East Side, being apparently broken-up ice, refrozen into tough solid masses, very unequal on the surface, and with deep overhanging edges, under which the sea was washing with a hollow dismal sound.

We were too distant at this time to make out whether or not the Sound was frozen across, but it may be believed it was not with uninterested eyes I looked in that direction, which, four years before, had been taken by those of whose welfare so many were now looking eagerly for tidings. I would fain have struck at once to the westward; however, there was nothing for it but to wait patiently. So I made up my mind to pass the next month in Pond's Bay [Pond Inlet] as I best could, the hope never leaving me that I might yet succeed, one way or another, in getting up Lancaster Sound.

On the ninth we were reaching in to Cape Byam Martin, the snow-capped peaks of the Byam Martin Mountains towering up beyond.... In the evening we found ourselves off Cape Graham Moore, the northern point of Pond's Bay. It had now fallen almost a dead calm. Every one on board was on the alert and in high spirits, for the whalers consider that if they get to Pond's Bay the first week in July, they are sure to fall in with a run of whales, and so secure a full ship. The ship at this time making scarcely headway through the water, the master was talking of sending the boats into the bay, to see if they could fall in with a fish or two. The deck was thronged by the eager crew, the older hands pointing out the well-remembered features of the bold coast before them, each rendered memorable in their eyes by the slaughter of some huge "nine," or "ten footer," in former years. In speaking of the size of a whale, they estimate it by the length of the longest laminae [slab] of whalebone.

The harpooners were all busy in their boats, examining their guns, harpoons, and lances; the attention of every one else was directed towards the bay, when the sudden cries of "A fish!" "A fish close astern!" "A mother and sucker!" caused a rush to the boats; in an instant a couple were manned, lowered, and after her. There she is—a large whale, with the calf sporting about, and but a short way astern; the deep roust [sic], and the spouting fountain of her blast, contrasting with the weaker and lower one of the calf. Ah! they are down—the quick eye of the mother has seen the boats, and she is off. The faces around me on deck begin to elongate, and their owners begin to think that it will prove but a "loose fall" after all. But, no; the harpooner in the headmost boat is a sharp fellow and an experienced one—he has marked which way the fish had "headed," and he is off after her, bending to his oar, and urging his men to do the same, until the boat seems to fly over the water. For twenty minutes they pull steadily on in the same direction. Now, see! the boatsteerer is pointing ahead; it is the calf that has risen to breathe—had the poor mother been by herself she would have been far enough by this time, but she stays by her heedless offspring, and she now appears at the surface also, within a "fair start" of

the boat. A few strong and steady strokes, and they are at her. "He's up! he has pushed out his oar; and stands to his gun." There is a puff of smoke; an instant afterwards a report—the boat is enveloped in spray, and the sea around broken into foam—as with an agonised throe the mighty creature dives, in the vain effort to escape. All this has been witnessed from the ship with the most breathless anxiety; but now every soul is bawling "A fall!" "A fall!" at the pitch of their voices, whilst the rest of the crew are tumbling *pell-mell* into the remaining boats, which are lowered almost by the run, and without the loss of a second, are off towards the "fast one," which is now seen, with its "jack" flying, a happy sight to the master, who directs it to be replied to, by hoisting the ship's "jack" at the mizzen. The harpooneers in the loose boats now station themselves around the fast one, but at some distance from it, to be ready to attack the whale the moment she appears at the surface, with the exception of one which remains beside it to "bend on," should the fish take out all its lines.

Half an hour is now past, and during that time the fish has been "heading" towards the ship, so that the boats are but a short distance from us. Every instant she may be expected to reappear at the surface. "There she is!" "Hurrah boys!" "She spouts blood." The first harpoon has been well aimed, and sent home with deadly force; she is already far spent; but a second and a third are sent crashing into her, and she dives again and again, but for a shorter space each time, until at last she lies almost motionless on the surface, whilst with the long and deadly lance they search out her most vital parts. "Back! back all of you! she's in her dying flurry." No, she is too far spent, it is only a faint flap of her heavy fin, and a weak lash of that tail which, an hour back, could have sent all the boats around her flying into splinters. She turns slowly over on her side, and then floats belly up, dead. "Three cheers, boys, for our first Pond's Bay fish: I'se warrant ye, she's eleven feet if she's an inch, and I'm sure she's no been that ill to kill," cries out some excited harpooner. The equally excited men replying by three cheers of triumph that make the blue bergs ring again.

But it must not be taken for granted that the whale is always so easily captured as this one was. It is often a work of severe labour, and almost always one of considerable risk; but the excitement of the sport is such, that this is scarcely thought of. It is but seldom now, however, that a whale can show much fight, in consequence of the deadly effects of the gun harpoons, which are now constantly used by all the ships. It may be easily conceived how much more efficacious these are than the old hand harpoons, particularly when well aimed, and at a good range. A smart harpooner, however, generally manages to get fast with his hand harpoon, as well as his gun, being thus doubly secure of his fish.

All were of course highly encouraged at this propitious beginning of the fishing, almost at the very instant of our reaching the ground. After "flensing" the whale, we proceeded in to the land ice, and there made fast.

WHILE THE *Advice* cruised in the vicinity of Pond Inlet in July 1849 and prepared to penetrate Lancaster Sound as soon as ice permitted, the search for Franklin steadily gained momentum across 3000 miles of arctic territory. In the Western Arctic boats from HMS *Plover* were about to set off eastward along the Alaskan coast towards the Mackenzie River. Further east on the mainland coast John Rae was attempting to extend the boat survey carried out the summer before, which had then searched from the Mackenzie River to the Coppermine. In Lancaster Sound the naval search vessels *Enterprise* and *Investigator*, under the command of Sir James Ross, had spent the winter at Port Leopold on the northeastern tip of Somerset Island and were waiting to get clear of the harbour ice and sail westward in Barrow Strait. Finally HMS *North Star*, quite unknown to Ross, was on her way from England bringing enough supplies to enable *Enterprise* and *Investigator* to remain out a second winter. In fact the supply ship failed to reach the expedition, which, after unsuccessful efforts to get further west, returned home in the autumn of 1849, ending one of the Admiralty's first search expeditions with not a trace found of Franklin and his men.

Interestingly, the Ross expedition had been close to solving the Franklin mystery. In May and June of 1849 two man-hauled sleds, one commanded by Ross himself and the other by Francis Leopold M'Clintock, had travelled down the west coast of Somerset Island. Although they could not know it they had chosen the same south-trending Peel Sound routeway taken by Franklin's ships *Erebus* and *Terror* in the summer of 1846. If Franklin had only taken the time to leave a few messages in well-marked cairns along the shore, telling that he had passed, arctic history would have taken a completely different course. The sled party, finding no trace of the missing ships, turned back for Port Leopold at 72°38′N, unaware that a year earlier and less than 200 miles further south, the 105 men still alive on the lost expedition had abandoned their ice-trapped ships, landed on King William Island, and begun the impossible task of dragging sleds and boats south towards fur trade posts a thousand miles away.

By a curious irony, M'Clintock, who with Ross had so narrowly missed reaching the scene of the disaster on this occasion in 1849, was destined to discover the grisly skeletons and relics littering the shores of King William Island ten years later, when he would command his own expedition sponsored by Sir John Franklin's widow and the British public. In the meantime the search would continue, nourished by false hopes, and hindered by the unforgiving arctic environment.

For Captain Penny and Robert Goodsir in the *Advice* in July 1849, there was no possibility of sailing west in Lancaster Sound because fast ice once again barred the way. But for the moment there was plenty of work at hand, for whales were now abundant off Pond Inlet.

For the next ten days we continued our fishing, with varying success, occasionally casting off from the ice, and running a short way to the southward, as the whales seemed to be more or less plentiful. We were more generally astir during the night than during the day, for it almost invariably happened that "a fall!" if called at all during the four-and-twenty hours, would be about midnight or after it; then adieu to sleep for the next eight hours at least. But there was little privation in this, for I think there are few men who having once seen the exciting scene of a whale hunt, would for an instant prefer their beds to the pleasure of seeing it again. For some days we had scarcely seen any fish. A small straggler would be seen occasionally, and was soon dispatched by some one or other of the ships; but still there was nothing like a "run"; and, although we ourselves were at that time better fished than our neighbours, yet we were not getting on half fast enough for some of the more impatient spirits. For my part, every successive capture we made was a sort of disappointment to me, for the more we got, the less chance was there of our getting up Lancaster Sound, my only aim and object. Still, it was pleasant to see all around me happy at every accession to the cargo, which was to take comfort and happiness to many a fireside and family during the winter, and for which all the poor fellows were toiling so hard. But, in spite of my so far selfish feeling, I am certain I was as keen and as eager as any one on board whenever the exciting cry of "A fish!" was heard, or the still more exciting and rousing one of "A fall!" and I managed more than once to be "in at the death," and take my share in the sport, as well as in the drenching shower-bath of hot and greasy blood.

It was late in the evening of a brilliantly clear and warm day—one of those days which but too seldom enliven this land of eternal ice and snow, and which, when they do happen, contrast so delightfully with the many days of dreary mist which the visitor of the arctic countries has to endure.

Two or three of the hands were lounging listlessly about the decks, all the watch being "on the bran" in the boats, stationed along the edge of the ice, to which the ship was made fast, and the rest of the crew sound asleep in their berths. The master had just gone up to the crow's nest, to take a look around him before turning in. He had not been there many minutes, before his quick and well-trained eye saw whales blowing beyond a point of ice some ten miles distant. The welcome news soon spread that the long-looked for "run" was at length in sight, and ere long every soul was astir and ready for the sport. The boats were immediately lowered, those in the "bran" were called along side, all their kegs filled with bread, beef, and water, and a small supply of grog given to each. The master was anxiously reiterating his orders to each of the harpooners; whilst some of the keenest of them were running up to the crow's-nest, and as they came down again were asserting that they saw the whales spouting like "steam-coaches, only far thicker." Most of the boats were now sent off to meet the "run," but in a short time the whales, showing no inclination to come further into the bay, the rest were dispatched also, with orders to pull right out to them. I had no idea of remaining by the now almost deserted ship at a distance from the scene, so I proposed to go in the last boat, and, as we were short of hands, I had no difficulty in getting my offer accepted. We had a long pull before us, but the anticipation of the sport, the delightful calm of the evening, and the beauty of the scene around us, shortened the distance wonderfully. Looking towards the land from which we were pulling, nothing could be more beautiful than the immense extent of high and mountainous coast that was stretched out before us, broken across, as it were, by the opening of the bay, the whole variegated in the most beautiful manner by the lichen-coloured rocks, and the brown patches of vegetation appearing above the ground-work of snow; whilst half-way down the black precipitous crags of the shore hung a long filmy riband of gauze-like mist, tinted with the most delicate crimson by the level rays of the midnight sun. The whole, too, seemingly so close at hand, that more than once throughout the past day I had caught myself wondering at my laziness in not stepping across the narrow boundary of ice which separated me from the shore, until I recollected that the apparent mile was nearly fifteen, and these fifteen rendered unsafe by the decayed state of the ice. Nothing could be more tantalizing than this apparent propinquity, for, after months of confinement to the greasy decks of a whaler, it would have been an unspeakable luxury to have set foot on shore again, or to have been able to pluck even the simplest moss or lichen of the scanty flora of the shore before us. The longing that a landsman has to be on shore again, after a tedious sea-voyage, may be easily conceived; but to sail for hundreds of miles without being able to land, within an apparent stone's throw of a coast – desolate it may be, but still rich in gloomy grandeur of scenery, – creates a longing which it may well be believed is much more intense.

But we are now drawing nigh the scene of action; we had for some time been meeting numerous shoals of narwhals *(Monodon monoceros)*, whose blasts every now and then startled us, as they are almost as loud as that of the whale.

We passed a Kirkcaldy vessel, the crew of which were busily engaged, and pulling onwards; we shortly came up to one of our own boats, which we found had succeeded in killing a large fish of ten or eleven feet bone: the fish was floating at the edge of the floe, and the boat's crew would fain have had ours to join them in the laborious and irksome task of hauling in their lines. But we had no idea of this when there was sport to participate in a little farther on: so, after a few minutes spent in asking questions, how many lines she had taken out, &c., all of which seem so interesting to the true whaler, we had regained breath, and pulled onwards. About three miles further on we found a second boat with her "jack" flying, denoting that she was fast. Passing close to this boat, we found that the fish was taking out line with great force and rapidity, and that the harpooner was rather doubtful as to his being "well fast" or not; that is to say, he was uncertain

whether his harpoon was securely inserted into the whale; he had fired at a long range just as the fish was going down. We pulled in the direction in which she was "heading," where the rest of the boats already were; before we got up to them, she had made her appearance at the surface; a second boat had got fast to her, and just in time, as she was seen to "loose" from the first. She did not take out much line from this boat, but remained away a considerably longer time than usual, greatly to our astonishment, until we found that she was "blowing" in some holes in the floe, a good distance from the edge of it. One of the harpooners immediately proceeded over the ice with a hand-harpoon, trailing the end of the line with him, assisted by part of his crew, and from the edge of the hole drove his weapon into the body of the poor whale; whilst some of the others following plied the bleeding wretch with their long lances, so that she was soon obliged to betake herself again to the open water outside the floe. Here more of her enemies were waiting, for our boat was immediately upon her, and a gun-harpoon was at once driven almost out of sight into her huge side, which was already bristling with weapons. Our boat was on her very back as she dived, with an unwieldy roll, which sent it surging gunwale under, taking the line whistling out for a score fathoms, until the harpooner, knowing she was pretty well exhausted, stopped her way, by taking three or four turns round the "bollard". But every few seconds she would make a start, drawing the boat almost head under, until the line was permitted to run out again, which, as it did so, made a grinding, burring noise, eating deep into the hard lignum vitae of the bollard, enveloping the harpooner in smoke, and causing the most distinct smell of burning, which was only prevented from actually taking place by the line-manager throwing water constantly on it.

Again she appeared at the surface, but far exhausted, still she made a strong fight for it, lashing about with her tail and fins in fury whenever she seemed to have regained

It was not uncommon for whaleboats to be capsized or destroyed by blows from the whale flukes or pulled underwater when harpooned whales dove. From the Orray Taft *sketchbook. (Kendall Whaling Museum)*

breath. It was no very pleasant sight to see her tail quivering high up in the air, within but a short distance of us, and coming down on the water with a loud sharp crack, like the report of a dozen rifles, and which had it alighted on any of our boats, had power sufficient to have converted their timbers into something very like lucifer matches. A few more lances soon settled her, and ere long she was rolling on her back. The usual cheers of triumph were given, and we had time to breathe and shake ourselves, for it may be believed we had not escaped the showers of spray which the defunct had sent about so liberally.

The water far around us was dyed with blood and covered with a thick pellicle of oil, upon which the mollies were as busy as they could be, whilst the edges of the ice, as far as we could see, were deeply crimsoned; and a hummock on the edge of the floe, beside which the final struggle had taken place, was from the summit downwards streaked with the black blood which the last few blasts of the dying monster had sent over it.

Much to our satisfaction, we had little line to pull in, so that we were soon ready for another victim. It must not be thought, however, that I have been all this time an idle spectator. If one wishes to partake in this sport he must also partake in the labour. The whaleboats are necessarily so constructed that they can only contain their proper crew. But as I was able to handle an oar, from former practice, I had no difficulty in finding a place in them, and so gaining a closer view of the scene. The labour was severe, as we had already pulled upwards of fifteen miles, and that at full stretch, as hard as we could lay to our oars; but this was scarcely thought of at the time. It was only now when the excitement was over that I thought of fatigue or felt it. I had luckily pitched my pea-jacket into the boat when we left the ship, as I had a sort of idea we might be some time away, so I now rolled it up, placed it on the gunwale of the boat, and stretching myself out on the "thwart," slept as soundly as ever I did in my life. My slumbers, however, did not last long, for it was scarcely according to rule that any one should sleep in the boats on

Far left: Harpoons were made of soft iron to bend rather than break after penetrating the whale. (Dundee Museums)

Left: Between the dangerous and thrilling events of a whaling cruise there were sometimes quiet interludes as in this scene of summer arctic midnight. (Dundee Museums)

fishing-ground. But I woke thoroughly refreshed, and we were again in full chase after the "fish."

We had two or three unsuccessful bursts after them, but failed in getting within striking distance. We saw one of the boats, however, a short way from us fire at a large fish, which, on receiving the harpoon, leapt almost clean out of the water, head first, displaying the greater part of its huge bulk against the sky, until we thought it was going to jump right on to the floe. Suddenly reversing itself, its tail was seen high over the boat, and so near that for an instant or two we breathlessly expected to hear the cry of agony from the poor fellows as they were crushed beneath it. But she dived sheer downwards, quite clear of the boat, towards which we now pulled quickly to render assistance, more excited, perhaps, by the narrow escape we had just witnessed than they were themselves. Distant as we were from the ship, and notwithstanding the hairbreadth escape they had just made, the joyous shout of "A fall!" was now raised, and the jack displayed. Just, however, as we reached it, the line which had for the few seconds since the fish had dived been running out with lightning speed, slackened, and the strain stopped. The harpooner looked blue, and began slowly hauling in, his crew assisting, with

long faces; for, be it remarked, each man in a "fast boat" gets half a crown and the harpooner half a guinea. We sat gravely by, condoling with them on having lost their fish. In a few minutes the harpoon appeared on the surface, and was hauled on board, with sundry maledictions from the *heathens* of the unlucky boat. The whale had wrenched herself loose by her sudden and active leap, for the massive iron shaft of the harpoon was bent and twisted upon itself as one would twist a piece of soft copper wire with a pair of pliers.

We pulled back again towards our former station. By this time we scarcely knew whether it was night or day. We had a sort of idea that we had been a night and a day away from the ship, but of that we were not certain. We had made repeated attacks upon the biscuits and canister of preserved meats, but although the appetites of steady-living people at home are pretty fair timekeepers, we found ours of little use in that way here.

I suspected it was again night, but I could scarcely think it possible, the time seemed to have passed so rapidly. But there was a *stillness* about the air that must have struck every one as peculiar to the dead hour of the night, and though I have noticed it in far different situations, it never struck me so forcibly as it did here. The light passing breezes and cats' paws which had dimpled the water for some hours back had died away. It was now so calm that a feather dropt from the hand fell plumb into the sea. But it was the dead stillness of the air which was so peculiar. No hum of insect, none of the other pleasant sounds which betoken it is day, and that Nature is awake, can be expected here even at midday in the height of summer, twenty miles from land, and that land far within the Arctic Circle, where, if one may say so, a third of the year is one long continuous day. Yet there is a most perceptible difference,–there is a stir in the air around,–a sort of *silent music* heard during day which is dumb during night. Is it not strange that the deep stillness of the dead hour of night should be as peculiar to the solitude of the icy seas as to the centre of the vast city? For many hours we lay quietly still, no fish coming near enough for us to attempt getting fast. But during the whole of this time they were pouring round the point of ice, and apparently running in towards the bay, almost in hundreds. The deep boom of their blowings, resounding through the still air, like the distant bellowing of a herd of bulls. My ear should have been pretty well accustomed now to the blast of the whales, but it was not until this time that I ever had noticed the peculiar hollow *boom* of their voice, if voice it may be called.

We thought at the time that the fish were running right into the bay, and imagined we could hear the distant sound of the guns, and the shouting of "falls" about the ships, which could just be seen. We were in no very good humour at the idea of not being in the thick of it, but we had no reason to complain as it turned out, for we learned, on our return, that the fish had never gone into the bay, and that scarcely any one had seen them on this occasion but ourselves. But we now had a good chance; a fish was seen beside the ice at no great distance from us, but beyond a "fair start." I have noticed a peculiarity about the whale,

that if there is a piece of ice within sight it will run towards it, and come to the surface beside it. And when beside a floe it always rises beside its edge, and never appears at any distance from it. And, moreover, if there should be a crack or bight in the floe, it is ten chances to one it will rise to blow in it, in preference to the outer edge of the floe. This is well known to the whalers. Such a crack being now opposite to us, and at such a distance from where the whale was last seen, it was likely she would rise there next, and we pulled towards it. Here we lay for some minutes in breathless expectation, our oars out of the water, and the harpooner silently motioning with his hand to the boat-steerer which way to "scull." Up in the very head of the crack the water was now seen to be circling and gurgling up, *"There's her eddy,"* quietly whispers our harpooner: *"A couple of strokes now, boys,–gently,–that'll do."* Looking over my shoulder, I could see first the crown, then the great black back of the unsuspecting whale, slowly emerge from the water, contrasting strangely with the bright white and blue of the ice on each side–then followed the indescribable hurstling [*sic*] roar of her blast. But short breathing time had she–for, with sure aim and single tug of his trigger-string, the keen iron was sent deep in behind her fin. *"Harden up, boys!"* he cries, and the boat is pulled right on to the whale, when he plunges the hand-harpoon deep into her back, with two hearty *digs*. The poor brute quivered throughout, and for a second or two lay almost motionless; then diving, and that with such rapidly increasing speed, that the line was whirled out of the boat like lightning. The usual signals were now made to the other boats that we were "fast."

For the first few minutes the lines were allowed to run out without interruption, then, one, two, three turns, were successively thrown round the "bollard." This had the effect of stopping her speed somewhat, but the line still ran out with a great strain. The boat's bow was forcibly pressed against the ice, and crushed through the under-washed ledge, to the solid floe beyond; the harpooner sitting upon his "thwart," allowing the lines to run through his hands, which were defended by thick mitts: stopping the progress of the fish as much as he could, as the rest of the boats were still some distance from us. Every few minutes the fish seeming to start off as with renewed strength, the boat's bow would be pulled downwards, threatening to pull us bodily under the floe. But then allowing the line to run out, the strain was partly removed, and the boat's head again rose, but only to be again dragged downwards. Upwards of twenty minutes had elapsed since we had "got fast," and the strain now began to slacken, but it was full time, – we were drawing nigh the "bitter end." The welcome sound of a gun was heard, and in a few seconds, looking down the edge of the floe we could see one of our boats with the well-known blue "jack" flying. A few fathoms more of line were rapidly drawn out, and then the strain as suddenly ceased. We commenced hauling them in, and whilst doing so, could see a third boat "get fast." The rest of the boats were now at hand, and as she appeared at the surface, closely surrounded her, and busily plied her with their lances. It was in about an hour and a half from the time we first struck her, that we heard the distant cheers announcing her death. From the time the second boat had got fast we had been busily engaged hauling in our lines, and thus slowly approaching the cluster of boats round the dying whale. But long ere we had finished this they had succeeded in killing her, and she was lying safe and sound, made fast to the edge of the floe. The boats now collected and prepared to tow the dead fish to the ship. This was even more tedious then hauling in the lines, but as I had volunteered to take my place in a boat, I said not a word, but tugged away at my oar in silence. Luckily, however, one or two fish were seen near us, in pursuit of which our boat and another cast off from those which were towing. The moment we were again in chase,

fatigue and languor vanished, and we stretched to our oars as heartily as we had done when we first left the ship.

We had a long, but a fruitless pull, and in the mean time a light breeze had sprung up, and we could see that the ship had "cast off" from the land ice in the bay, and was working down towards the boats and dead fish. We pulled towards her at once, and I was not a little glad to be able to stretch myself on deck again, after nearly forty-eight hours confinement to the thwart of a boat. A hearty welcome from the captain, who was not a little astonished to find me so fresh after my labours, and the tempting sight of smoking beefsteaks and *early potatoes* on the cabin table, soon made me all right, nor did I feel half so fatigued as I might have expected, and was later than even my usual time of retiring to my narrow berth in the little closet off the cabin, which was by courtesy termed the *doctor's stateroom*.

Two or three days after this, I had another opportunity of closely witnessing the death of a whale. She had been struck in a crack but a short distance from the ship. All the crew, except the "watch," who were on the "bran," were sound asleep in their berths below, fatigued after some days' hard labour. It is a most laughable scene to see a "fall called" under such circumstances. The one or two hands, who were walking quietly and gently on deck a second before, in order not to disturb the fatigued men below, are now seen dancing and jumping like madmen, on the half-deck hatch, screaming "a fall!" as if for their lives. The more active men of the crew are on deck in an instant, with ready bundle of clothes in hands, and shoes or boots slipped loosely on their feet. But it is generally a race who will be first into their boats, clothed or unclothed, and nothing is more common than to see half a dozen fellows rushing to the boats with nothing on but their woollen under-clothing, the rest in a bundle under their arm, trusting to the first stoppage to complete their toilette,

The cry of "A fall!" brought half-dressed men flying from the forecastle into the whaleboats, a pandemonium similar to this sketch of sailors turning out to reef sails. From the log of the Tarquin, 1862-63. (Peabody Museum of Salem, Massachusetts)

such as it is. Rather a sudden change this from their close and crowded "bunks" (as they call them) in the half-deck, to an atmosphere often far below zero. But neither the old whaling sailor, nor the green Orkney boy, ever seemed to feel it.

The stern boat was the only one now left on board. The master ordering it to be lowered, and getting into it himself, I jumped in with him. We pulled up to the "fast boat," to see how things were getting on, and found they were only fast with the gun harpoon, and not very well with that. Whilst talking to the harpooner of this boat, we heard a commotion amongst the others, and almost before

we had time to turn, bang! went one of their guns, and the fish was made almost secure. She seemed to dive under the floe, and reappeared almost at the same place, for she next came up within a very short distance of where she was first struck, when a third boat got fast to her, and before she dived again she was mortally lanced. When she next appeared at the surface, it was close to our boat; we were at her in a minute, when the ready lance of the master was twice buried deep behind her fin. She made a rush forwards, which pulled the lance out of his hand, but he soon had a second—we "hardened up" to the fish, when he plunged it into her side. She had been quiet enough

Fourteen years after Franklin and his men vanished, a search party breaks open a cairn on King William Island containing a brief message left by the expedition. (Public Archives Canada, C-3468)

hitherto, but it was now full time for him to cry, "Back, men, for your lives!" I heard a sudden whizzing, whistling sound in the air—I thought a black cloud has passed between us and the sun—a drenching shower of spray passed over us, and there was a loud *thud* upon the water on the other side of the boat, as her huge tail descended into the sea, which it continued to lash into seething foam for more than five minutes. I may be believed that whilst this was going on we all kept at a safe distance. It was, however, only the last struggle—"the dying flurry," and the huge mass was soon lying powerless and motionless before us. This was a female whale, and one of the largest we had yet seen.

IN JULY, WHILE THE *Advice* pursued whales off Pond Inlet, the first long-awaited news of the Franklin expedition was unexpectedly provided by Eskimos to the captain of the Kirkcaldy whaler *Chieftain*. It was absolutely dazzling; the *Erebus* and *Terror* had been in Prince Regent Inlet for three winters and were still there, with everyone alive and well! They were not far from Ross' ships *Enterprise* and *Investigator*, and the two expeditions were in contact. This oral report and a chart drawn by the Eskimo informant, who claimed to have been on board all four vessels, had been passed on to Captain Parker of the *Truelove* of Hull, and through him had reached Penny and the *Advice*. Parker considered the report reliable; Penny and Goodsir were sceptical. But who could say?

On 2 August the *Advice* left Pond Inlet, heading at last for Lancaster Sound and Prince Regent Inlet. The weather was clear, and eager eyes strained ahead, struggling with the optical illusions created by light refraction as they looked for a ship or a sign of human residence ashore. On the fourth the weather deteriorated; a rising gale created rough seas and forced the *Advice* to heave to under shortened sail. She was able to proceed westward again on the fifth, but the renewed spirit of excitement and expectation among the crew was short-lived, for in a short time they came up against a barrier of fast ice extending from north to south across Lancaster Sound, which prevented any further progress towards Barrow Strait or Prince Regent Inlet. Reluctantly, Captain Penny came about and headed back towards Baffin Bay.

On the way past the mouth of Navy Board Inlet the men constructed a message cairn on an island, and then they continued southward along the coast of Baffin Island pursuing whales whenever possible, and speaking a number of other vessels from time to time. Finding the mouth of Cumberland Sound blocked with ice, Penny ended the whaling and started homeward across the North Atlantic to deliver the impressive season's catch of 159 tons of oil and almost twenty tons of baleen, from seventeen whales. In addition to their shares in the proceeds of this impressive cargo the crew would later receive part of an award of £1,000 for services in the Franklin search.

As it turned out, the scepticism of Penny and Goodsir concerning the Eskimo report had been entirely justified. There was no truth whatever to the story. Born out of linguistic misunderstanding, the use of leading questions during interrogation, and the native tendency to supply answers that would please the white men, what appeared to be authentic news of the Franklin expedition had then been exaggerated and embellished as it passed from ship to ship. It was news that everyone desperately wanted to hear, and it travelled fast. When the *Advice* arrived at last in Scotland, and Goodsir asked the harbour pilot if there was any news of Franklin, the man assured him that the expedition was safe. The crew rejoiced, but as the pilot provided more details Goodsir recognized the unmistakable signs of the same "rascally Eskimo report" they had heard at Pond Inlet. It had beaten them across the Atlantic.

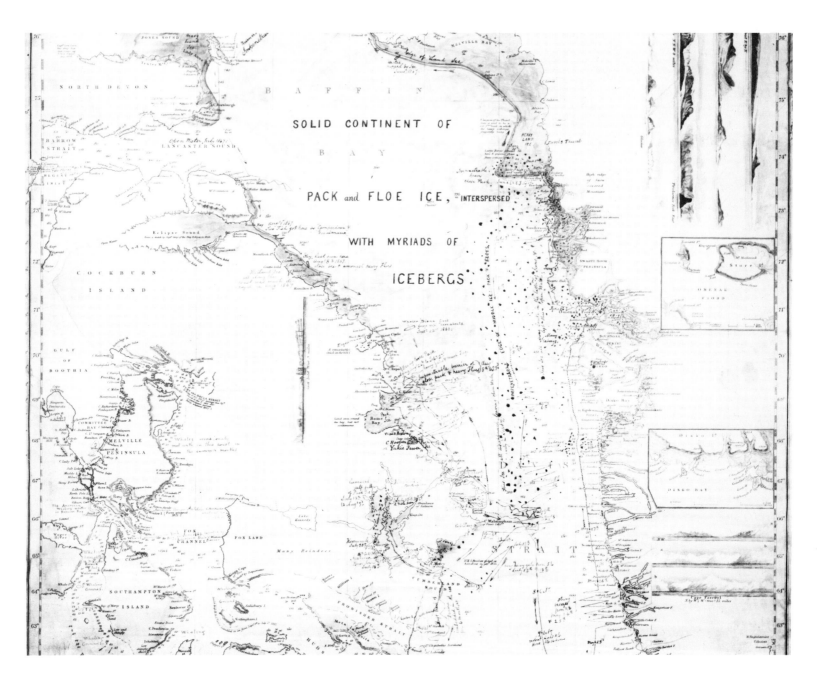

This detail from a whaleman's chart demon-strates the cartographical eccentricities confronted by nineteenth-century arctic navigators. The route shown is that of the whaler Ravenscraig *in 1867. (Kircaldy Museum, Scotland)*

Nine

To Form a Sort of Settlement

They were in a most miserable state, being reduced to perfect skeletons.
STEWART A. LITHGOW, 1853

WHEN DAVIS STRAIT WHALING expanded to the western regions of Baffin Bay in about 1820 whalemen had unknowingly committed themselves to a different set of environmental conditions, which would come to modify their whaling routine and introduce new navigational and survival risks. The climate along the Baffin Island coast was more severe; the set of currents and of floating ice was southward rather than northward; the duration of pack ice and landfast ice was greater; and the fiords and harbours of the coastline were less accessible and little known. Although West Side whaling was to prove more productive (for a few decades at least), the higher yields were achieved only at the cost of longer voyages and adaptation to more demanding physical conditions.

After crossing from Melville Bay in early July the whaling vessels would congregate around the mouth of Pond Inlet. When the run of whales came to an end off Pond Inlet some captains would sail westward into Lancaster Sound, while others would begin cruising down the rugged coast of Baffin Island towards Cape Dyer, almost 600 miles to the southeast. On this deeply fiorded littoral, lofty mountains rose several thousand feet directly out of the sea, supporting ice caps and valley glaciers. The scenery, when it was not obscured by fog or precipitation, was majestic and inspiring – even to those in trouble – but no one would have considered the coast a hospitable one. Ashore one could appreciate the splendid views and admire the hardy, ground-hugging plants, but to European eyes there seemed to be precious little to support human life.

Nor was the situation at sea one to inspire confidence. Charts of the mid-nineteenth century showed the rocky coast miles out of position, provided little or no information on water depths, tides, or currents, failed to depict the interior parts of most fiords, and contained some fantastic geographical embellishments based entirely on conjecture, such as waterways extending from Baffin Bay across Baffin Island to Foxe Basin. In addition to inaccurate and misleading charts whaling masters had to contend with enormous compass variations. In Davis Strait and Baffin Bay their magnetic compasses pointed almost a right angle away from true north. While it is possible to determine how magnetic variation differs from region to region and from one year to the next, magnetic surveys had not yet provided the necessary information at this time. The masters therefore had to depend largely on visual navigation and the sum of their collective experience, working from landmark to landmark as best they could, usually in company with one or more other vessels.

Early and late in the season, ice was a constant danger. Stories of the disastrous ice drifts of earlier years were never quite forgotten. And for a ship beset in the pack there was another awful possibility. Icebergs, responding to deep currents, sometimes moved in directions contrary to the drift of the shallower, wind-driven floes of pack ice. A ship could be carried helplessly on a collision course with an iceberg, or held fast while a berg crashed irresistibly through the floes towards the vessel. This perilous situation, which had been experienced by the *Dundee* in 1826 (chapter 2) and the *Viewforth* in 1835 (chapter 5), would also, as we shall see, confront the *Alexander* in 1853.

The ship *Alexander* from Dundee was one of four Scottish vessels whaling in Davis Strait in 1853. Thirteen Hull vessels comprised the remainder of the British fleet, in addition to which there were two American whalers from New London, the *Amaret* and *Georgiana*. These latter represented the beginning of a gradual increase in American participation. After constituting an important part of the fishery during the pre-independence years of the eighteenth century, the Americans had shifted their attention elsewhere, and none of their whaleships had been seen in Davis Strait until 1846, when the *McLellan* had initiated a new era of American enterprise in the Eastern Arctic. By the mid-sixties American whalers would be almost as numerous as British.

The author of the selection below was a third-year medical student at Edinburgh University, Stewart A. Lithgow (in later life Major-General Lithgow, of Fanhope, CB, DSO, AMS). What induced the young man to ship out on a Dundee whaler in the summer of 1853 is not explicitly known, but such ventures were almost traditional in the medical school of the University, which doubtless provided a substantial number of surgeons for the Scottish whaling fleet through the years. Lithgow boarded the *Alexander* at Earl Grey Dock, Dundee, on 28 March. Hundreds of relatives and well-wishers cheered as the ship was towed out to an anchorage. The crew was mustered, but it was discovered that twelve men were missing. (A last-minute change of heart was by no means unusual on arctic voyages.) Lithgow found it hard to fit into his cramped quarters. There was no place to stow his kit, and on retiring he found that the length of his body exceeded the length of his berth by about twelve inches. The vertical dimensions were no more generous; every time he attempted to sit up in his bunk he rapped his head firmly. And if that was not enough his cabin-mate snored "a dozen trumpets from nasal organs."

The frustrations of adapting to the shortened dimensions on board ship, and to the lack of privacy, affected others as well. On the following day a seaman leaped overboard, swam to a fisherman's boat, and made good his escape. It must have appeared to Lithgow that the attitude of the whalemen was not characterized by an appropriate amount of enthusiasm or eagerness to get under way. At last, on 30 March 1853, the *Alexander*, in company with the *Advice, Heroine,* and *Princess Charlotte*, weighed anchor and set sail for the whaling grounds of Davis Strait.

More than three and a half months later, on 19 July, the *Alexander* arrived off Pond Inlet. Few whales were to be seen so Captain Sturrock took the vessel slowly southward in company with several other ships during the remainder of the month and through August. Pack ice was still plentiful along the Baffin Island coast and the ships spent most of their time either moored to pieces of ice or "dodging" back and forth in small areas of open water. The ice impeded the movements of the ships and often frustrated attempts to get at the "rocknosing" whales close inshore with the whaleboats. On the other hand the slow progress along the coast gave the men (and Lithgow) occasional opportunities to take brief leave of the cramped quarters on board ship and stretch their legs on the arctic tundra. The whalers reached Home Bay about 21 August, and by 6 September were approaching Broughton Island.

Whaling surgeons were sometimes able to stretch their legs ashore, collect specimens of plants, birds, and rocks, and admire the view from a nearby summit. This is the steam whaler Maud *in 1889, photographed by Livingstone-Learmonth. (Public Archives Canada, C-88296)*

TUESDAY 6 SEPTEMBER

Light winds and cloudy weather. At 3.45 A.M. called all hands and sent four boats in shore in search of fish. Went along with one of them, to see what like the land was. About an hour's pulling brought us to the opening of an inlet of, as near as I could estimate, a mile and a half in breadth.... The hills rise very abruptly on every side to a great height, and have a very barren aspect, as they are merely bare rocks, the vegetation, what there is of it, being chiefly confined to the lowest parts, and here even it is not always found, as it is only on some spots where the soil is more genial, that some mosses and diminutive looking shrubs, rarely exceeding four or five inches in height, can be seen. Having pulled for about a mile up this inlet, landed at a small beach, where we found a seal's skin, very recently flinched, that had been left by some of the natives. Upon opening a small cairn close by it, discovered pieces of the flesh and part of the intestines filled with blubber, and tied at the ends with whalebone. As this was evidently a stock that they had laid past, we covered them up and left them as we found them. After this set off with one of the crew to the top of one of the mountains, apparently about two thousand feet in height above the level of the sea. After some trouble owing to the difficult nature of the footing in ascending, at last got to the summit, but not before it had turned quite thick and begun to snow. But having come so far we did not wish to leave before having seen something, so we resolved to wait a little to see if it would clear up. Nor were we disappointed, for in less than half an hour the thickness all cleared away and then there was a view presented before us which certainly amply repaid our patience. The sea and the land were spaced out just like a map and the ships appeared so small that one might think they were possessed by Lilliputians. The prospect altogether was very fine. Returned to the boat much pleased with the short excursion, having been absent between three and four hours. By a quarter to six in the evening was aboard the ship, having seen no fish....

Left: Eskimos spearing fish on fast ice at Eclipse Sound; a steam whaler in the background. (National Library of Scotland)

Near right: Eskimo skin tents incorporating wooden poles and doors probably obtained from whalers or wrecks. (Dundee Museums)

WEDNESDAY 7 SEPTEMBER

Moderate breezes and cloudy weather. At 4 A.M. called all hands and sent four boats in shore in search of fish. Two and sometimes three boats on the watch near the ship, a number of fish having been seen. In the afternoon a very large one rose close beside the ship. The stern boat – the only one left – was lowered, and just as the harpooner was up, to fire, off went the fish. At 8 P.M. all the boats returned. They had seen a number of fish, and one of them had got fast to a fish supposed to be a loose one but it afterwards appeared that one of the *Advice*'s boats was fast to the same fish. ...

THURSDAY 8 SEPTEMBER

At 4 A.M. called all hands and sent five boats in towards the land to fish. Went into one of them along with the captain as they were going to a place where the boats had seen natives yesterday. In about three hours we came to the place where they were located. Upon coming to the beach a young fellow of about seventeen or eighteen years of age

came to meet us. He could speak a little English and seemed to be rather intelligent. The encampment consisted of three tents, situated on a low level point of land, close by the water. Each tent or hut was composed of seals' skins sewed together and supported in the form of a pyramid by pieces of wood, broken oars &c. got from some ships. The interior was very warm, and the floor was strewed with an awful confusion of seal and deer skins, pieces of seal flesh, half-picked bones, blubber &c., the stench arising from which on your first entrance is almost intolerable. In the midst of this mess sits the "kuna" with her legs in something the same position as our "knights of the needle" when on duty. The inhabitants consisted of two middle-aged women, one very old one, seven or eight children, and the beforementioned young man. The men were away farther north, at the sealing. The women were stout and rosy-cheeked, but certainly anything but good looking; their feet and hands, however, were singularly small, and well formed. They chatted away to us in their own language with great vivacity, and seemed to be very

Above: A large umiak *off the coast of Greenland, rowed by women and steered by a man. (Dundee Museums)*

Above: A Baffin Island man, Shangoya, and his wives, one of whom wears clothes of Euro-American style and fabric. Metal implements, weapons, and containers, as well as wood and cloth articles, were popular trade items from the 1820s on. Photo by F. N. Gillies. (Dundee Museums)

A group aboard the Scottish whaler Diana *at
Dexterity Fiord, Baffin Island, in the 1890s.
Photo by F. N. Gillies. (Dundee Museums)*

Scurvy and exposure claimed the lives of many whalemen, whose crude graves are still to be found along the coast of Greenland and Baffin Island. (National Library of Scotland)

happy and contented. Their lamp consisted of a hollow stone set up on end, with some oil in it. What they used for a wick, was stuff something like cotton, that they obtain from a plant, which exists in considerable quantities in some spots of the land. The whole of them were dressed in skins, the females' upper garment bearing a ridiculous resemblance to a long-tailed coat behind, only the tails were rather on an extensive scale. They had their faces and hands all tattooed, and this seemed only to be peculiar to the female sex, at least so far as I had an opportunity of judging.

After having my curiosity satisfied, returned to the boat, and set off to see and fall in with some fish. Had not long left when we saw one and immediately set off as fast as we could pull towards it, but it had disappeared before we got near the place. Upon rounding a piece of ice came sud-denly in view of four immense sea horses lying fast asleep on it. The captain was unwilling to fire, in case of the whale coming up again, so we backed the oars to keep off for a little, and see if it would come, but the noise caused by the oars wakened them, and off they floundered into the water, with an awful splash. When we had lain for an hour or two longer got the boat's sail put up, and set away out the inlet with a nice breeze. Took it down at the mouth of this and pulled in towards the south cheek of it (Cape Broughton) where we landed. On rambling over the point fell in with the last resting place of a sailor, who ten years ago had found a grave on this lonely spot; a board over his head had an inscription carved out upon it with a knife, and although exposed to the action of the elements for full ten years was still as fresh and legible as the day on which it had been executed. Got on board the ship about 5 P.M....

FRIDAY 9 SEPTEMBER

At 4 A.M. called all hands and sent four boats in shore to fish. Noon, strong breezes with snow. The boats returned; no fish seen. In the afternoon proceeding southward among streams of ice....

MONDAY 12 SEPTEMBER

First part of the day, light winds and cloudy weather. At 8 A.M. set all possible sail, the ice coming in upon the land and the young ice making so fast, that the ship was nearly two hours under canvas before she would move a foot. Latter part, strong breezes and thick with snow, the vessel plying among a number of bergs and heavy pieces of ice from which arises great danger during the darkness of night, especially when it is thick also, this preventing their being seen before you are close at them. About midnight passed so close to a large berg that one might have touched it with an oar. Everybody was in the greatest consternation until we got past it, as the ship was drawing closer to it every minute, and there was another large one upon the other side so that we were in great hazard of losing the ship....

TUESDAY 13 SEPTEMBER

Strong gales and thick with snow, the vessel dodging and beating to windward among a number of icebergs and pieces of ice. Midnight, wind increasing; close-reefed the topsails....

WEDNESDAY 14 SEPTEMBER

Strong gales and thick with snow. At 4 A.M. wind still increasing; stowed the mizzen and mizzen topsail. At daylight, it clearing a little, discovered that we had got into Merchants Bay a considerable distance, and that we were fast driving down upon the lee shore. Put up some more sail, as much as the ship could stagger under, to get her rid of the danger by going off to the eastward. Eight A.M.

stowed the foretopsail, the vessel in among a great number of bergs and large pieces of ice, under close-reefed foresail and maintopsail. Night, still blowing as hard as ever, and so thick that it is impossible to see more than a ship's length ahead. Some of the men on the forecastle, one on the bowsprit, and another on the foretop, to look out, while others are stationed at the various ropes, ready to trim the sails on the shortest signal, the wind blowing so hard that one could barely hear another speak. This is certainly the most fearful night that ever I spent in my life....

THURSDAY 15 SEPTEMBER

The wind and snow as bad as ever, if possible. At 3.40 A.M. there arose from the forecastle the awful cry of "an iceberg on the lee bow," which threw every one into the greatest alarm. All hands were called, to give them a chance of their lives, at least that they might not be drowned in their beds, as we expected every second that the ship would be dashed to pieces. The few minutes of suspense that followed this cry I don't think I will ever forget, as if our anticipations had been realized. Not a soul of us could have escaped unless by a miracle, as the great sea running at the time would prevent the boats being of any avail for our safety, so that it was with feelings of the greatest joy that I heard the second mate exclaim that she would clear it yet, and when he was fairly past it the congratulations at our narrow escape were general. We were so close to it that the wash of the waves coming off it, was flying over our ship's sides. I suppose never did any crew wish more sincerely for the appearance of daylight than what we did on this occasion, the first symptoms being hailed by everyone with the greatest joy. When day at length broke, although thick with snow, it disclosed to our view a perfect forest of bergs that we had driven through...and any person to look at them one would have thought it next to impossible to have come the way we did in such a gale, without losing the ship.

Ran the ship about ten miles to the southward, to get clear of bergs. At 7 P.M. hauled the ship by the wind under close-reefed topsails, as a chain of immense bergs were seen to the southward. Midnight, more moderate and cloudy; set all possible sail, and plied northward....

THE YEAR 1853 marked some important changes in whaling procedure. They evolved out of a lingering reaction to the disastrous events of the 1820s and 1830s (chapters 2,5,6). Several very bad ice years had then vividly demonstrated the vulnerability of ships to pack ice pressure, and the helpless situation of crews whose vessels became frozen in for several winter months. They could not be reached by ships from Britain because they were surrounded by impenetrable ice, and they could not seek assistance from whaling bases or European settlements on the Baffin Island coast because none existed. Several persons, among them the arctic explorer James Clark Ross, R.N., had advocated the establishment of one or more shore stations at Davis Strait, as early as 1837. Year-round bases on the West Side, while offering emergency refuge for ships and crews, would also be advantageous for whaling. By employing whaleboats directly from shore in late fall and early spring, they could extend the whaling season and reduce the dependence upon European ships. Only a vessel or two would be needed each summer to collect produce from the stations and deliver supplies for the next winter.

While this idea simmered in cautious British minds an American whaleman, Captain Quayle of New London, Connecticut, implemented a different sort of modification to the traditional system of one-season voyages. In the fall of 1851 he left a group of volunteers from the whaler *McLellan*, under mate Sydney Buddington, to winter ashore in huts in Cumberland Sound. (George Tyson, one of the group, recounts his experience of this winter in chapter 12.) The men kept healthy on native foods and emerged in 1852 in fine shape with seventeen whales

to their credit, an achievement that gave the wintering concept a powerful boost.

The Scottish whaleman William Penny saw this American initiative as a threat to assumed British sovereignty in the region and in 1852, as a countermove, he requested from the British Colonial Office, on behalf of his Aberdeen company, a grant of land encompassing the whole of Cumberland Sound, for the purpose of establishing a permanent whaling colony.

The proposal was strongly resented and successfully opposed, not by the Americans, but by whaleship owners in other British ports. When their petitions blocked Penny's project he took a different tack, taking two Aberdeen whalers, the *Lady Franklin* and the *Sophia* (the latter under George Brown) on a wintering voyage to Cumberland Sound in 1853. The crews were to live on board the ships, rather than ashore as the American whalemen had done two years before.

One of Penny's opponents in the land grant matter was a Hull merchant, John Bowlby. He had claimed that to give Penny's company exclusive rights in Cumberland Sound would be to destroy his own plan – already in advanced stages – for setting up a whaling station there. While Penny persevered in seeking official authorization Bowlby went right ahead without it. In June 1853 he dispatched the three schooners *Bee*, *Seaflower*, and *Wellington*, with all the materials, provisions, whaling gear, and personnel necessary for the founding of a permanent shore base in Cumberland Sound.

Thus in the summer of 1853, while Stewart Lithgow accompanied the whaler *Alexander* on a traditional one-season voyage, two ships from Aberdeen and three from Hull were converging on Cumberland Sound, the former planning to winter and the latter intending to erect a whaling station. Both these techniques were new in the whaling.

Penny was a man of great energy and capable leadership and he had already wintered in the Arctic in 1850–51, while employed by the Admiralty on a Franklin search expedition. He turned this experience to good account in 1853–54. His two ships secured a total of 39 whales, yielding almost 230 tons of oil and more than 15 tons of baleen.

The Bowlby project, on the other hand, appears to have failed completely. The events are veiled in obscurity but some revealing details emerge coincidentally from the narrative of Stewart Lithgow. The *Alexander*, while sailing into Exeter Bay in mid-September, encountered a small, derelict schooner, disabled by ice, drifting helplessly with her small powerless crew towards certain destruction. She was the schooner *Wellington* – one of Bowlby's ships.

FRIDAY 16 SEPTEMBER

Moderate winds and clear weather. Set all sail and plied into Exeter Bay. Four boats set inshore, in search of fish. At 2 P.M, while standing into the bay, discovered a vessel in the bottom of it with her ensign union down (the signal of distress). At 3 P.M. came to an anchor in twenty fathoms water. At the same time a boat was manned and sent on board the stranger to see what was the matter. Found her to be the schooner *Wellington* of Hull, who had left that port on the seventeenth of June, laden with a general cargo and bound for Kemmisook, a place on this side of the straits about a hundred miles to the southward of this, where they were going along with other two schooners to form a sort of settlement for the purpose of prosecuting the whale fishery &c., their engagement extending over the space of [three?] years. They had only six weeks' provisions on board when they left, and owing to adverse circumstances, their voyage had been protracted to nearly fourteen weeks, for six of which they had been enduring the greatest hardship for want of food, their only diet for all that time consisting of barely a gill of rice each per day. When we picked them up they had only nine days provision at this rate left them. They were in an extremely weak condition when the boat boarded them, being quite unable to manage the ship, so that our men had to take her in

charge, and brought her alongside. The schooner had run foul of a berg, which had carried away the covering board and five stanchions, and stove in part of the decks, so that she was making a deal of water, and not fit for any person remaining in. Took the crew on board, consisting of five men and a boy. They were in a most miserable state, being reduced to perfect skeletons; two of them indeed could not have lived more than a day or two longer, they were so much exhausted....

SATURDAY 17 SEPTEMBER

Moderate breezes and variable winds. Four boats out on the watch for fish, the rest of the people on board getting water from the land and unrigging the *Wellington*....Set fire to the schooner in the afternoon, which burned till midnight. Was ashore in the afternoon for some time.

MONDAY 19 SEPTEMBER

Fresh breezes with showers of snow. At 4 A.M. called all hands and sent four boats out to fish, others employed getting fresh water on board. At 5.15 P.M. the boats returned, one fish seen. The vessel still at anchor....Went ashore today and fell in with a sailor's grave. From the "headboard" saw that he had been one of the *Joseph Green*'s crew, and had died there in 1847.

TUESDAY 20 SEPTEMBER

Light winds and variable weather. At 4 A.M. called all hands and sent four boats out in search of fish, the rest of the people getting the ship made all ready for sea. A great quantity of ice setting into the bay from the northward; the wind being light could not proceed out. At 5.30 P.M. the boats returned, one fish seen....

WEDNESDAY 21 SEPTEMBER

At 4 A.M. called all hands. At 6 A.M. weighed the anchor and sent six boats to tow the ship out to sea. A great deal of ice setting into the bay so that we had some difficulty in getting out. Noon strong breezes and cloudy. Got into a large hole of water, set all possible sail and hoisted the boats up....

A RCTIC WHALESHIPS were normally between 200 and 400 tons, with hulls specially strengthened inside the bows and doubled outside the waterline. They carried crews of forty to fifty men.

The transport *Wellington*, by comparison, was a diminutive schooner of only twenty-five tons, manned by "five men and a boy." This tiny vessel and crew had not been up to the task of coping with arctic ice, and sending them from Hull to Baffin Island seems nothing less than folly. While the intention of setting up a whaling base was perfectly reasonable, the expedition outfitted by the merchant Bowlby appears to have been badly planned, inadequately equipped, and poorly executed – a case of gross mismanagement with near-tragic results. The ships were too small and had probably not been modified for navigation in ice. The food stores for the station, instead of being divided among the three vessels, were placed only upon the *Bee* and *Seaflower*, so that the *Wellington* carried no part of them, only lumber for the buildings. The food on board the ship was insufficient for an arctic voyage, and the crew were forced to make six weeks' provisions last for more than thirteen weeks. There were no firearms on board to facilitate hunting for meat. The fact that they were two hundred miles north of their destination in Cumberland Sound suggests an incompetent or inexperienced master. After the vessel was damaged by collision with an iceberg during a gale, the starving men could do little more than keep her afloat. Their time was running out. If they

had not been sighted by the *Alexander* they would soon have perished, and the explanation of their disappearance would have vanished with them.

On 22 September the *Alexander* manoeuvred among a great number of bergs and floes and on the following day fell in with the whalers *Advice* and *Venerable* about 45 miles southwest of Exeter Bay. A consultation of the three captains took place, "the result of which was," Lithgow recorded, "that they could not remain longer in the country with safety, and therefore they came to the conclusion to bear up for home, at which I am very glad."

Captain Sturrock turned all hands to removing whaling gear from the boats and securing everything on deck for sea. Two days later the *Alexander* was driving homeward at speeds of six to nine and a half knots, registering almost 200 nautical miles in some twenty-four hour periods, and steadily drawing away from her consorts *Advice* and *Venerable*. The ship arrived at Dundee on 18 October with the blubber from three whales.

Stewart Lithgow wrote in his journal that the arctic experience on board the *Alexander* had been "on the whole one of a very pleasant description, and to which I shall always look back with pleasure." It was now time to resume life ashore and turn his attention once again to academic pursuits. He shaved off his beard for shore-going, and went to collect his pay. Wages and "fish money" for the voyage came to £14.12, an amount double the sum required for his lodgings during the entire university year (which may go some way towards explaining the tendency of Scottish medical students to ship aboard whalers). The summer vacation was over; more important considerations were now at hand. On the first of November his terse diary entry reads "College opened today. On the hunt for lodgings."

This depiction of "Life in the Forecastle," from
J. Ross Browne's Etchings of a Whaling
Cruise, *suggests some of the shipboard antics*
of crews during idle moments. (Metro Toronto
Library)

Ten

Snug for the Winter

May guardian angels prove your steps attend
And guide you safe unto your journey's end;
In every stage of life may you most happy be,
And when far distant offtimes think of me.
MARY JANE WHITEHOUSE to her departing husband,
May 1859

Sent the crow's nest down. Making the ship ready for the winter.
Blowing strong at all times. All hands dancing....
ALBERT JOHNSON WHITEHOUSE in Cumberland Sound,
September 1859

ALTHOUGH THE ATTEMPT BY HULL SHIPOWNERS in 1853 to establish a whaling station failed, the practice of wintering on board ship was successfully initiated in that year. The two Scottish ships under captains Penny and Brown, and two American whalers, both commanded by Buddingtons, took up winter quarters on the southwest shore of Cumberland Sound and remained frozen into harbour ice for the next ten months. The experiment showed the advantage of being present when the whales made their reappearance in the spring. Penny's crews killed almost twice as many whales from the floe edge in May and June 1854 as they had in the previous summer and all four whalers returned with large cargoes of oil and bone. The events of 1853–54 also proved that fears of spending a winter in the Arctic had been exaggerated; if ships were adequately outfitted and securely positioned, and their crews were competently led, the arctic winter held no terrors. Wintering quickly became an accepted whaling technique and in almost every year during the next few decades several whalers could be found frozen in at Niantelik, Kekerten, or other harbours in Cumberland Sound, their crews idly passing the time until the resumption of whaling in May. Wintering was not universally adopted, however; most whaling captains continued to make one-season voyages.

When whaling masters intended to winter they considered the safety of their vessel first and then the welfare of the crews. They required a harbour large enough to accommodate several ships (for whalers seldom wintered alone) but sheltered from strong winds and heavy seas, a place with reasonable depths for anchoring, a good holding bottom, and a stable ice surface in winter. If the shore was mountainous they looked for nearby areas of low relief where they could get ashore easily to store supplies, travel inland to hunt, and obtain access to fresh water. The presence of an Eskimo encampment in the area was considered an added advantage.

Having selected a suitable harbour the captains would bring their vessels to anchor, usually in mid-September, and wait for harbour ice to form. When the ice was stable, and strong enough to hold the ships in position, they would take the anchors up through holes in the ice and hoist them on board for the winter. Once held firmly in the harbour ice the ship became the men's base, a familiar environment amid the hostile arctic surroundings, a protective wooden shell in wintry blasts, a storehouse against starvation, a working place, dormitory, and recreational space. The ship provided security but not what we would today call comfort. The crew's quarters in the forecastle were crowded and although coal stoves were used for heating they were employed sparingly, and getting small volumes of heated air from one or two stoves to remote parts of the vessel was always a problem.

Eskimos came to play a vital role in the successful wintering of whaling ships. They had already developed close ties with the whalemen, and were often hired to assist in the summer whaling. Through employment and trade they had obtained guns, ammunition, used whaleboats, whaling gear, and many other useful articles; and because their appetite for manufactured goods tended to increase rather than diminish over the years, they normally congregated near wintering whaleships and offered their services as "ships' natives" in order to acquire the items they needed or desired. Their principal contribution during the winter months was to provide the whalemen with fresh meat—mainly caribou—which not only offset the monotony of shipboard provisions but also prevented scurvy if consumed regularly. Aside from their invaluable services as hunters, Eskimo men worked transporting blubber to the ships by dogsled, guiding captains and mates on hunting excursions, and participating in spring whaling. Their women often made fur clothing and sleeping bags for the crews.

The Hull whaler *Emma* was one of five British whalers taking up winter quarters near Niantelik in Cumberland Sound in the fall of 1859. The ship departed from Hull on 30 May, and arrived off Durban on 21 July. Three days later, while cruising off Cape Hooper, a lookout noticed nine men on the floe edge. They turned out to be from the Dundee whaler *Advice*, which had been wrecked the day before. Captain Simpson took the men on board and sent twenty of his own crew across twelve miles of fast ice to see what might be saved from the wreck and to locate the rest of the survivors. Evidently, the subversive effects of salvaged spirits had undermined the resolve of the

Sailing into Winter quarters

majority of the shipwrecked crew and convinced them to remain at the vessel rather than to seek assistance. The same strong drink now proceeded to weaken the resolve and sap the energy of their would-be rescuers as well. After excessive imbibing, a disorganized band of more than seventy men began to straggle back across the ice in carefree confusion, like an itinerant cocktail party. This irresponsible frolic was to have a tragic outcome, for quite unexpectedly the temperature dropped and the wind rose. Most of the men managed to reach the floe edge and were evacuated with difficulty to the *Emma* in the gale-lashed seas, but the most inebriated—some of whom were by now incapable of walking—had to spend most of the night on the ice. By morning three men were dead and several badly frostbitten.

Carrying fifty men from the *Advice* as well as her own crew of forty-five, the *Emma* worked southeastward along the coast of Baffin Island, gradually distributing the survivors to other ships. Finally unencumbered, the ship then proceeded south in company with the steam tender *Isabel*, entered Cumberland Sound on 12 September and dropped anchor in a harbour (un-named) already occupied by two American and one British whaler. Three more British ships came in the next day.

Auxiliary steam propulsion was only in the early stages of

Far left: American whalers—two barks and one schooner—taking up winter quarters by way of a channel sawed through the harbour ice. Sketch by Timothy Packard. (Houghton Library)

Left: Ships frozen into position in winter harbour. Sketch by Timothy Packard. (Houghton Library)

application in the British whaling fleet but the *Isabel*, one of the first experimental vessels, used her engine to advantage in the forthcoming weeks, towing other ships in and out of harbour, exploring adjacent parts of the coast in search of a good winter harbour, ranging further afield with her own whaleboats and crews, helping the *Emma* move to another harbour on the 23rd, and later assisting her consort to shift to deeper water within the anchorage. The two ships moored side by side, while somewhere in the vicinity, within a few hours sled travel, the *Sophia* and *Union* were also preparing for winter.

The crews of the *Isabel* and *Emma* dismantled crow's nests, topgallant masts, and yards, and stored coal and spare provisions ashore. They obtained fresh water from streams before freeze-up, and later sledded slabs of newly formed pond ice back to the ship to stack for melting down as needed through the winter months. The cooks butchered two pigs brought out from England on the hoof, while the coopers assembled shakes [barrel staves] into casks to hold blubber, and closed in upper deck spaces to provide more room for work and recreation.

In November a film of bay ice spread out steadily from shore, growing thicker day by day. On the eighteenth some of the men ventured to walk gingerly upon the elastic surface, and two days later, although only "half a fist" thick, the ice was supporting a number of sailors gaily gambolling about and even experimenting with dog sleds. This convenient platform was soon thick enough to bear traffic safely and hold the ships firmly. Anchors were hoisted, and cut through the ice. The ships had become, effectively, extensions of the land surface.

While the crews kept busy during October and November making the preparations for winter, they did not neglect the whaling. They built a lookout hut on an outer island out of stones and turf, with a stove inside for heating, erected a flagpole for signalling the appearance of whales, and cruised regularly in the whaleboats, venturing forth in freezing temperatures and sometimes in snow flurries. Only high winds and large seas appear to have kept the small boats shorebound on these damp and bitterly cold autumn days.

A number of Eskimos from the settlement at the harbour participated in the various activities. Some had been hired as ships' natives to assist in the autumn and spring whaling, and hunt for meat during the winter. Others, including the women, played a role more social than economic, coming on board frequently for dances. On the fifth of November, the last day on which the boats went out after whales, the men of the *Emma* burned an "hifigy" of "Guy Fox." The winter social season then swung into action at the harbour.

The excerpts below are taken from a daily record of events written by Albert Johnson Whitehouse, a boatsteerer on the *Emma*. He reveals that although whaling had to be suspended from November to May the wintering whalemen were not entirely idle during this time. He reports a variety of outdoor activities and sports on the harbour ice (even on the coldest days), frequent games of cards, checkers, and dominoes on board ship, regular visits and many overnight stays by Eskimos, and a surfeit of dancing, singing, and drinking. There is no indication of boredom; the arctic winter appears to have been all "fine fun" for Whitehouse and his shipmates. But on the other side of the coin are references to frostbite, drunkenness, fights, theft, complaints, protests, and near-mutiny, which make one wonder about the quality of the leadership on board and the influence of such undisciplined behaviour upon the Baffin Island natives.

Whitehouse's journal is a tiny book about three by five inches. On each page he managed to crowd almost 300 words, all written minutely in pencil and now faded and smudged. The challenges of deciphering this indistinct midget text are considerable even with the aid of a magnifying glass, and the difficulty of the task is compounded by the unorthodoxy of the diarist's spelling, grammar, and punctuation. In pack ice the ship was "in a dangers posian." After a death, "the sail maker sowed the man up in Canvis and threw im over bord with out Buril servis." An event was "remarkhubel." Some of the crew "been at the warter…seed sevarl fish." Ships' names appear to be in code: *Piyeneer* (*Pioneer*); *Agerstaner* (*Agostina*); *Georey Any* (*Georgiana*); *Annable* (*Hannibal*); *Chieftan* of Kalkardy (*Chieftain* of Kirkcaldy). In addition, Whitehouse was frustratingly vague about locations, so we cannot know exactly where the various ships wintered, where the spring whaling base was located, how far it was to the fresh water ponds, or other important details. There are problems even when he supplied local placenames such as Emma's Island, Penny's Harbour, Mallemuk Head, and Yankee Point, because these names, although familiar to the whalemen of those days, subsequently fell out of use before anyone had consigned them officially to maps or gazeteers. Despite these imperfections it is worth including some of Whitehouse's diary because it is one of the earliest surviving narratives of an intentional wintering on board a Davis Strait whaler and it provides many details of life at winter harbour.

SUNDAY 11 DECEMBER

Twelve noon, master of the *Sophia* on board and master of the *Union* on board, and some of the crews of both ships. Some of our men away shooting. Seen several hares and foxes but we have not shot any yet. Weather fine through out. Cook and engineer of the *Isabel* the same [i.e., sick]. All hands pretty merry. All the Yacks on board &c.

MONDAY 12 DECEMBER

Twelve noon, weather very stormy. Been employed all day in making mats &c. Cook better and the engineer a great deal better. Night, all dancing and singing, some with their faces blacked singing negro songs. All the Yacks on board. Very strong [wind?] night &c.

TUESDAY 13 DECEMBER

Twelve noon, heavy gale of wind from the northeast. Very heavy gales of wind last night from the south-southwest. Called all hands at 10 P.M., the canvas and mizzen and main staysail all blowing away, everything blowing adrift about the ship. All hands employed most of the night. Today all been employed most of the day in mending the sails &c. Night, blowing heavy from the northeast, the ice breaking up outside. All hands dancing and singing. All the Yacks on board &c. Snow. …

THURSDAY 15 DECEMBER

Twelve noon, weather fine through out. Employed fetching ice for ship's use and rigging shears on board of the schooner [*Isabel*] ready for to take her mast out tomorrow

Men hauling blocks of pond ice to a wintering whaler in Hudson Bay in 1864-65; it will be used for drinking water. The house on the right is probably a storehouse. From the Orray Taft *sketchbook. (Kendall Whaling Museum)*

if it be fine weather. We are going to brig-rig her. Night, dark and cloudy. All dancing and singing. All the Yacks living on board &c.

FRIDAY 16 DECEMBER

Twelve noon, weather fine through out. All been employed in various duties. Spread the canvas over the decks again. Called all hands at 7.30 A.M. and none of us would turn out. Master came and called all hands himself &c. All hands called on deck to receive a lecture. We have all to walk round the capstan every morning at 8 A.M., or forfeit two days' pay, or stop our provisions, but that will have to be looked into. The tradesmen all employed in making materials for rigging the schooner. All the Yacks on board, several from the other ships, men and women. Fine night. Dancing and singing.

SATURDAY 17 DECEMBER....

Twelve noon, all hands employed in lifting the schooner's mast. The warp broke so we left it for another day. Twelve noon all hands on the ice firing rockets out of one of the

boats. Fired several into a mass of ice and they exploded sending the mass in all directions. The last one we fired burst the gun but no one was hurt. We lost the boat that went to the *Sophia* some time since; we lost her in the last gale of wind we had. The ice broke up at the *Sophia* and she lost four of her own boats; it was a heavy one. Several fresh Yacks came today from Kingaway [Kingua]. We have got about fifty men women and children now. Weather fine but very cold. Night, all dancing and singing. All well on board of both ships....

TUESDAY 20 DECEMBER

Twelve noon weather fine but freezing very keen. All hands been employed all day in fetching ice from one of the large bergs in the sledges with all the dogs yoked.... Several Yacks on board from the other ships. Night, fine. All dancing and passing the time away &c....

FRIDAY 23 DECEMBER

All employed lifting the schooner's mast out and other duties. Weather fine, snow at times. Some of the *Union's*

Football on the ice, photographed in 1894 by A. M. Roger. (Dundee Museums)

men on board. We are going to have a cricket match with them on the fourth of January if all goes well. The Yacks shot several seals &c.

SATURDAY 24 DECEMBER: CHRISTMAS EVE

Weather fine with snow at times. All sorts of games going on. Night, all hands went aft and sung the masters a song.... Some drunk &c....

SUNDAY 25 DECEMBER: CHRISTMAS DAY

Weather very cold and snow. All hands on deck at 8 A.M. and got dram. Some away shooting, and others in bed. We are going to keep the day up tomorrow. Night, fine. All hands in good spirits &c.

MONDAY 26 DECEMBER

Weather very cold, snow at times. Twelve noon, all hands dining in the 'tween decks. We had twelve bottles of brandy mixed up for toddy, four plum puddings, half of a pig and brandy sauce, and many other things. Four P.M. all dancing and singing in the 'tween decks, some drunk &c. Kept it up till twelve midnight, both masters among the crew. All kept from quarreling. Fine [fun?] &c.

TUESDAY 27 DECEMBER

Weather very cold and freezing very [keen?]. Several of us got slightly frostbitten. Most of our crew gone on board of the other ships, some shooting &c. *Union*'s men playing at cricket. Some of our men on board of the *Sophia* tonight. Picked our boat up that we lost in the last gale. Got the lines all safe but the boat is stove. Very cold &c. Two frostbit &c.

WEDNESDAY 28 DECEMBER

Freezing – thirty-nine degrees below zero – with snow at times. Some of the *Union*'s men on board.... Master of the *Sophia* drinking very heavy.

SATURDAY 31 DECEMBER

Weather fine. Several men ill with frostbitten feet. All the Yacks on board. Night, all hands went aft in the cabin singing on board of both ships, singing the old year out and the new in. Midnight, mutiny on board, all hands falling out on the half deck and 'tween decks…but Monday morning [will] have all hands on deck. Most of them in the half-deck drunk, some forward &c.

SUNDAY 1 JANUARY 1860: NEW YEAR

Weather fine. Some of the *Union*'s men on board today. Some of our men on board of theirs. Some away shooting. Saw several partridges and two foxes but got none. Some of the men's feet very bad with frost.… We have plenty of falling out on board of both ships. Fine night.

MONDAY 2 JANUARY

Blowing strong from the north-northeast with snow and heavy frost. All hands on deck today about Saturday night's racket. [Illegible word] to put the first man in irons that begins again, whether he belongs forward or aft. I have been before the committee today myself. Several fresh Yacks came today. Some of the *Union* men on board. They have been having a dinner today and getting drunk. I have cleared myself.…

WEDNESDAY 4 JANUARY

Called all hands at 6 A.M. and left the ship at 8 A.M. for the cricket match. The *Sophia*'s crew won. *Union*'s crew was fourteen and *Sophia*'s nineteen, and after that all three ships' companies ran a race for a sovereign. One of our men won it. We all went on board of the *Sophia* after the game was over. Some stopped all night and others came on board of our own ship. I got on board at twelve midnight. Thirty-nine degrees frost [below] zero. Master came on board the same time. We did not have such a game as we expected—far from it. Two or three got drunk. All the Yacks were there. Several of us got severely frost bit—face and feet, &c.

THURSDAY 5 JANUARY

Weather fine with heavy snow all day. All our men came back today. Cook got himself bad with frost. We have no work to do till Monday. All the Yacks on board and my cat

died about two or three o'clock this morning with cramp &c.

FRIDAY 6 JANUARY

Weather fine. Caught two foxes yesterday and today killed one and kept the other alive in a cage. We have got four or five traps set. All the Yacks on board tonight singing and dancing. Snow at times. One of the *Sophia*'s men on board &c.

MONDAY 9 JANUARY

Frost forty degrees below zero. All hands employed in fitting the….bunks all round and plaiting sinnet. The bunks are that cold that we can't sleep in them. We have to turn out two or three times in the night to warm ourselves. Knocked the bulkhead down yesterday so as the galley fire will warm the bunks. Built a bulkhead up yesterday across the 'tween decks to make the ship warmer. Night all hands pretty merry, some playing at cards. Some of the men's feet is very badly frost bitten…and scurvy &c.

WEDNESDAY 11 JANUARY

Morning fine. All hands playing at cricket and rounders. Some on the land. All the Yacks away seal shooting. Several hands ill, some rather touched with the scurvy. Night blowing very hard from the north-northeast. Playing at cards and checkers &c.…

SATURDAY 14 JANUARY

Weather fine. All hands playing at cricket and rounders. Night, we had a Yack's wedding in the cabin. Most of the Yacks drunk but we have not a chance to get drunk forward. Playing at cards and checkers. We ended the night with a fight &c.

SUNDAY 15 JANUARY

Weather fine but very cold. All hands running about on the ice. Several got frostbit. Several fresh Yacks came on board. The fox is living yet. He eats first rate. We trucked several partridges from the Yacks today. We had prayers in the cabin this morning &c.

MONDAY 16 JANUARY

Weather fine. All hands employed in fetching ice on board. Twelve noon, a sledge came from the *Union* with two men to fetch our doctor on board. One of their men—the carpenter—fell overboard last week and got severely frostbitten in the feet and legs. The doctors cut five of his toes off—two off one foot, three off the other, and his big toe off from his left foot—and they expect to have to cut the other off, if not his feet. Our cook's foot is very bad. We expect that his feet will have to be cut off yet. Night, all singing, playing cards and checkers. Several fresh Yacks came on board and some of our Yacks gone away to Kingtong seal fishing &c....

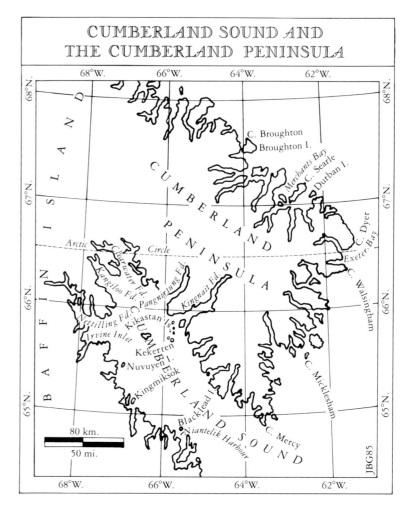

THE WINTER CONTINUED, with the crew of the *Emma* fighting it all the way. They hunted partridge, shot at targets and an occasional wolf, trapped foxes, and played cricket and rounders with abandon, whatever the weather, cheerfully accepting the consequences of discomfort and frostbite. But as a reminder of the real business of the expedition several whales put in an early appearance on 27 February just beyond the floe edge, about eight miles from the ship. Within a week the crews of the Scottish ships *Union* and *Sophia*, wintering nearby, had sighted seventeen whales, and on the thirtieth of March the men of the *Emma* saw three. In April the whaleboats were cleaned up and the lines dried and coiled in preparation for the spring whaling.

Being on hand for the floe edge whaling in early May was the principal advantage of a wintering voyage. Ships sailing from American or British ports to the whaling grounds would at this time be on the Greenland coast, prevented from making their way over to the West Side by a few hundred miles of impenetrable pack ice. The Greenland whales, smaller and more manoeuvrable than the sailing vessels that pursued them, and capable of swimming beneath the ice for a few miles at a time, were much more adept at moving through the pack, and almost invariably reached the Baffin Island coast well ahead of the ships. At Cumberland Sound whales were sometimes sighted so early (27 February in 1860) as to suggest the possibility that they wintered within the pack ice, either in or near the mouth of the Sound. Whether migrant or resident, they were likely to be found in the Sound in late winter or early spring, and only native hunters or over-wintering crews could get at them.

In winter the upper, or northwestern, part of Cumberland

Right: Native komatiks *and dog teams were useful during spring whaling at the floe edge when blubber had to be transported back to the frozen-in ships for making off. This sketch by Timothy Packard appears to be based on an illustration in one of Sir Edward Parry's journals. (Houghton Library)*

The Expedition to the North

Sound was covered with fast ice extending from shore to shore. But the lower, southeast, part of the Sound contained drifting ice floes among which whales could find enough open water to surface and breathe. It was part of their instinctive pattern to move up into Cumberland Sound during the summer as the fast ice melted but they could penetrate no further than the floe edge at any particular time. Whaling crews at Blacklead Island and Niantelik were within reach of the open water and therefore in a good position to exploit the whales as they congregated at the floe edge in May and June.

The procedure was to move the whaleboats, camping equipment, and provisions from the ship harbour to the floe edge itself, or to a promontory or island nearby. This was accomplished on *komatiks*, Eskimo sleds, drawn by large teams of dogs. The crews would reside there in tents, keep lookouts on watch, and launch their boats from the fast ice to pursue whales. When the wind blew the pack in tightly against the floe edge or the weather turned bad, they would haul their boats out of the water and wait for a change. Dead whales were towed to the floe edge and flensed in the water alongside, then the blubber and baleen had to be sledded back to the ship, still frozen into harbour ice.

A half dozen years before the voyage of the *Emma*, when

Captain William Penny's ships *Lady Franklin* and *Sophia* had been among the first whalers ever to winter intentionally in the region, they had experienced an unusually severe season in which the floe edge stood twenty-one miles from Niantelik. With three crews at the ice edge killing whales Penny had as many as twenty-two Eskimo sleds working steadily to transport the bulky blubber of seventeen whales back to harbour; the total distance travelled in this shuttle service amounted to 14,000 miles! In May 1860, however, the floe edge was only eight miles away—a more typical position—and fewer whales were killed, so that the logistical problems were less challenging. A camp was established on an island close to the open water, probably Blacklead Island, later the site of one of the most important whaling stations in the Arctic and the first permanent mission established on Baffin Island.

THURSDAY 12 APRIL

Weather fine, wind north by west. All hands employed in taking casks to island and put a tent up to live in when the fishing comes on. All the Yacks away sealing. Some of Penny's Yacks on board. Night fine, wind north by west. Some washing their clothes, playing cards &c.

FRIDAY 13 APRIL

Weather fine but very cold. Frost twenty-three degrees below zero. All hands employed in fetching ice. Twelve men all employed in coiling the boats' lines. Coiled four boats' lines—speksioneer's, loose harpooner's, and our boat's lines, and mate's of the *Isabel*. Both masters and mate fit out today. Several Yacks belonging to the other ships on board. Night several of our Yacks came on board, brought several seals. Some playing cards, dominoes &c.

SATURDAY 14 APRIL

Weather fine. All hands employed in taking boats to the island and various other duties. Several Yacks on board from the other ships. Some of our natives came tonight. All hands singing and dancing, some drunk &c. Edgar Grasby been frightened tonight—saw something on the ice but we don't know what. Some drunk. Master and mate not friends &c.

MONDAY 16 APRIL

Weather fine, wind north. All been employed in taking two boats and seven casks. Some of the natives watching the island for to see that all is safe, others away sealing. Night strong wind from the northwest. Seen several fish. Some [cording?] washing &c.

TUESDAY 17 APRIL

Weather cold, wind north. All employed in fitting both ships out. Sent up the *Isabel*'s topgallant mast and yard and set up the [small?] topmast, rigging, &c. Sent our topgallant rigging up &c....

FRIDAY 20 APRIL

Weather fine. All hands employed in fetching shakes from off the land, and ice from the berg. Got [eleven?] sledge loads. Twelve noon, all hands employed spanning harpoons in &c. Cooper setting up shakes. Carpenters away at

boats. Night, some of our Yacks came from the water, brought several seals. Weather fine. All hands well. Some playing cards and checkers, washing &c. Good Friday—hotcross buns tonight.

SATURDAY 21 APRIL

Weather fine. All hands employed in various duties, clearing all the snow from the ship. Most of our Yacks came tonight, brought a good few seals with them &c. Masters and several more been at the water today. They saw two fish &c. Night fine, wind northeast. Some playing cards, checkers, dominoes, &c. Some of the natives drunk and some of the crew....

MONDAY 23 APRIL

Weather fine. All employed in various duties. Our master been at the *Union* about them getting some of our natives and dogs. Both masters have had a row &c. Mate and bos'n fell out on the ice. Second mate and one of the prentices [apprentices?] been fighting. So we have been in a frap all day. Mate has seen several fish and unicorns. Night fine, [wind] northwest. All hands well. Some mending clothes, washing &c., cards, play game.

WEDNESDAY 25 APRIL

Weather fine. All hands employed in taking two boats to the island. Took second mate's boat to the Yankey Island and the other to the *Emma*'s Island. Twelve noon all employed in coiling boats' lines....Night, wind north by east with heavy snow. Several more of our natives came. They have brought a good few seals &c. One fox dead. Lost one out of the cask today. Dogs after him but no use &c. Dancing &c.

THURSDAY 26 APRIL

Weather fine, wind north. All hands employed in taking two boats to the Yankey Island. Twelve noon all employed among boats' lines &c. The Yacks brought eighty-five seals, young and old, last night. Fine breeze from the northeast. All hands in good health, some washing, mending &c.

FRIDAY 27 APRIL

Strong wind from the northwest with heavy snow. All hands employed in various duties. Several Yacks on board. Night strong wind and snow from the northwest. All hands well, some washing, playing cards, &c. Me washing cabin.

SUNDAY 29 APRIL

Weather fine, wind west by south. Me and two more men been to the water today. We saw nothing but two or three seals, the ice running very fast and [crushing?] up. Several more natives came today. Two snow igloos built on the ice close to the ship and three in the island for them to live in. Night wind southwest. Me and another been fighting tonight. We are going to finish it tomorrow. Mate of the *Sophia* got three fish at Newboyen [Nuvuyen Island], one six feet three inches bone, another five feet, the other three feet three inches. We expect to begin tomorrow &c.

MONDAY 30 APRIL

Strong wind from the southwest with heavy snow. All employed in clearing the ship of snow &c. Night strong wind from the northeast. All the Yacks on board. All dancing and singing. No shaving allowed on board on the first of May &c.

TUESDAY 1 MAY

Weather fine, wind northwest. All hands employed in making drogues for the natives' boats to drogue fish with. Several more Yacks came today. We put the garland up early this morning. We have two mess pots &c. on the strength of it. There has been a row on board of the *Sophia*. The men has took charge of the rum puncheon. The *Union* [she?] saw one of the men give one of the

natives called Bulleygar two black eyes so he wants to leave and come here, but we don't know how it will [illegible word]. Master of the *Isabel* has been at the *Sophia* to [illegible word] about our harpoon gun but the master was drunk and would not come up out of the cabin. He was [treated?] like a dog. Night fine. All hands well. All the natives on board &c.

WEDNESDAY 2 MAY

Weather fine, wind north. All hands employed in fetching ice from the berg. The *Isabel* fetching ice. Afternoon all employed in cleaning dirt from the ship's side and scattering it about the floe for it to eat through the ice so as the ship will be easier to get out. Sledge and dogs gone to Newboyen for to fetch some more Yacks, another to Kingnait for the same purpose. Night fine. All the Yacks on board &c. ...

SUNDAY 6 MAY

Weather fine, wind southwest. All hands in good health. Brown from the *Union* has been here with four of his men like roaring lions for five dogs that he says he has trucked from the Cape Searle Yack. Night all the Yacks praying and singing in their way, and cursing Brown &c. We have got orders for the floe edge in morning. Saw two fish today &c.

MONDAY 7 MAY

Weather fine, wind south. Called all hands at 6 A.M. and took two boats to the floe edge—loose harpooner's and a Yack's boat, all the Yacks helping us. Night fine, wind south. Very little water to be seen. Two or three Yacks came from Kingaway, brought ten seals with them. Another fox dead. All hands well &c. ...

WEDNESDAY 8 MAY

Weather fine, wind northeast. All hands employed amongst bos'n's work. Master been out to the water with

Capture of three Bears in Cumberland Inlet.

sledge and dogs, saw one fish close to the boats, shot one seal. Night fine, wind northeast. All natives on board. Sledge came from Newboyen – three Yacks came. They brought three seals with them. They have got six fish at Newboyen and three bears. All hands well except their eyes and some are nearly blind with being on the ice....

FRIDAY 11 MAY

First part of twenty-four hours wind southwest with heavy showers of rain. All the Yacks on board, two or three of

them ill. All our men well in health. Night, wind south and heavy snow showers. Some dancing, playing checkers, singing &c. Steward playing concertina &c....

MONDAY 14 MAY

Weather fine, wind north. Called all hands at 7 A.M. and all hands went to the floe edge. Put two boats in the water. We had to take one of them out again to get repaired – she made too much water. We broke two sledges with the hummocky ice – we had a very bad road. The ship is

Left: Three polar bears captured by the crew
of the American whaler Andrews in
Cumberland Sound in 1865-66. Sketch by
Timothy Packard. (Houghton Library)

Right: Narwhals were often killed by whaling
crews for the tusks. (National Library of
Scotland)

about six miles from the water. We saw three fish and a good few unicorns and white whales. A good few seals—shot two or three. The *Union* has got two bears in the water. The Yacks are left in charge of the boats tonight—our turn tomorrow night &c. All the natives on board. We are all well but most of us got pretty wet for the ice is very bad &c.

TUESDAY 15 MAY

Weather fine, wind north. The Yacks got fast to a fish last night at 10.30 P.M. All hands flensing today. The fish three feet four inches bone, a she-fish. All the dogs dragging blubber on the [illegible word], me and two more boats' crews at the floe edge all night. Saw one fish during the day but a great many white whales. All the *Union*'s boats fishing....

THURSDAY 17 MAY

Weather fine, wind northeast. Called all hands at twelve midnight. The ice broke up and the boats and crews drove

about five miles down the gulf. We got them on safely with taking the lines out of them.... Most of our men ice-blind, me and both masters in the bargain.

FRIDAY 18 MAY

Snowy, wet weather, wind southwest. All the natives away with sledges and dogs dragging the blubber to the island. Both masters there looking after the natives, some of them blind. Most of our men a good deal better. Afternoon all hands away to the island making the fish off. She has filled about four tons. Night fine, wind south. Most of the natives inland. Some of the *Sophia*'s natives here. Fine fun going on &c., some of the Yacks drunk....

TUESDAY 22 MAY

Strong wind from the south-southwest. All hands employed taking casks to the island and rigging tent for us to live in. All the Yacks gone to live at the island. Afternoon filling water off the floe. There is plenty on the ice now. Night wind north-northeast and fine. All the

Left: Modern remains of a whalemen's graveyard at Niantelik, Cumberland Sound. (Gil Ross)

Right: Wooden headboard of a whaleman's grave at Niantelik. (Gil Ross)

natives on board. Me very ill, been forced to knock off work &c.

WEDNESDAY 23 MAY

Weather fine. Called all hands at 4 A.M. and all went away to the floe edge. The *Sophia* got a fish. She killed the master and drowned another man. Several are hurt. The boat capsized. Picked the master up before he sunk. Fish four feet bone. The *Union* got another fish between six feet one half [*sic*]. Night all hands stopping on the island. All the Yacks living there.

THURSDAY 24 MAY

All hands [two illegible words] away at 4 A.M. Wind southwest. Two boats' crews away [after] fish, others away launching the boats across the ice. We hauled boats out of the water tonight. We saw one fish today. Night, most all hands sleeping on the island. Me and three more came to the ship tonight to sleep. We have got very poor lodgings on the land. All the natives on the land &c. All hands got wet today.

SUNDAY 27 MAY

Weather fine, wind south. Some of our crew and the *Union* got another fish early this morning—six or seven feet [bone]. Our mate and boatswain on the *Isabel* all night. Very little water to be seen today. Most of our men at chapel in the cabin. We have prayers every Sunday. Night fine. Some of our men and the *Isabel*'s gone to island to sleep tonight. Strong wind from the southwest. Tesuwin

IN. ERECTED
TO THE MEMORY
WM. McKENZIE
WHO DIED APRIL
25.1861 AGED
18 YEARS A NATIVE
OF THE B.C. ALERT

come across from Niantelik today with three or four boats and crews, for to fish for the bone. We expect that we shall get the blubber but we don't know exactly yet, for he is on Penny's Island. Some of the natives on board &c. Twelve months today.

MONDAY 28 MAY

Strong wind from the southwest. Called all hands at 5 A.M. All hands employed in dragging both the waist boats down to Penny's Island. Tesuwin is going to fish for us and his crew. Night fine, breeze from the south. Natives on board. One of our men got the scurvy. Some ill &c. White stocking day.

TUESDAY 29 MAY

Weather fine, wind southwest. Called all hands at 6 A.M. All employed in dragging casks to the island. All the natives employed with sledge and dogs dragging casks. Night fine, wind southeast. Some of our men sleeping at the island. Little water to be seen. Natives away fishing &c. ...

FRIDAY 1 JUNE

Weather fine, wind northwest. Called all hands at 3 A.M. and put six boats in the water, three natives and three of them manned with our crew. We saw nothing but unicorns and white whales and a good many seals. The Yacks shot several. Relieved boats at twelve noon. Night hauled them out of the water. Wind from the south and fine. ...

MONDAY 4 JUNE

Weather fine, wind north. All employed in various duties. Very thick and rain at times. Some of our men at the island. Night, wind northwest. All hands at the island. We lost ourselves several times between the ship and the island owing to it being so thick. A good deal of water to be seen. All the other ships' boats away. Our natives away in the boats. ...

WEDNESDAY 6 JUNE

Strong wind from the south with rain at times. All hands at the island all night and day. No boats away today. Tesuwin the native got two white whales today. ...

THURSDAY 7 JUNE

Weather fine, wind south. Called all hands at 5 A.M. No boats away today. The natives sealing. They got five. Night fine, wind south. All hands at the island. Very little water to be seen. We put one boat in the water after a dead fish's crang. We thought it was a fish.

FRIDAY 8 JUNE

Wind south. All hands on the island. No boats away. Tesuwin shot another white whale today. We got the blubber. Some of our men ill at the island. ... Me and the

steward come to the ship tonight for our sugar but we did not stay. Two coopers along with us. Some of the natives away deer [caribou] hunting. They saw a great many but only got one of them. The cabin got it—we did not get any of it &c....

MONDAY 11 JUNE

Weather fine, wind northwest. Called all hands at 4 A.M. All been employed in sawing both the ships adrift. We have got the *Isabel* moved about four ship's lengths today. We have had a good deal of blasting with gun powder. We have used about thirty pounds of powder today. Three of our men bad with the scurvy and two more ill. A good few fish been seen today and yesterday. Night all the natives on board. They have been sealing—got five or six. The ice

about five feet thick in some places. Most of us that have been on the ice got overboard today &c.

THURSDAY 14 JUNE

Weather fine, wind south. Very little water, the ice closing very fast. Four A.M. all hands flensing one of the fish at the floe edge. She is six feet bone. We got relieved from the other fish and brought a sledge load of blubber to the island. All the other hands dragging blubber to the island. Night wind north. Master and the natives making the fish off. [Officers?] getting rest for to start again. Tesuwin got a fish.

FRIDAY 15 JUNE

Weather fine, wind south. All hands employed in rolling some more casks on the island. Some of the natives away at the boats. Very little water. Some of our men very bad in the scurvy. Cook very nearly dead last night with the scurvy. Night fine, wind south. The ice slacked at 8 P.M. and we got her (the fish) into the floe edge and flensed her. She is five feet bone, three tons of blubber &c....

SUNDAY 17 JUNE

Strong wind from the southeast with heavy snow and rain showers, thick at times. The ice breaking up in a great many places....

TUESDAY 19 JUNE

Strong wind from the south with heavy rain and thick at times. All the natives on the ice shooting and getting fast to white whales and unicorns. Water close in to the rocks. It is lined with small fish. We got two unicorns and three white whales....

WEDNESDAY 20 JUNE

Fine breeze from the south with rain at times. All hands making off blubber. Steward at the ship for the provisions. All the natives in the ice after white whales and unicorns. They got one white whale. Night fine, light wind from the northeast. Me and two more men put the steward on board for to fetch the grub in the morning. One of the native women laid in on Sunday but the young one is dead. Several of our dogs got the distemper and died. Our cook out of his bed. Several bad in the scurvy &c....

THURSDAY 21 JUNE

...Me and the steward and two more men away at the ship. We have been away all night. We shot four dovekies and wounded several. The men that have got the scurvy are a good deal better &c.

FRIDAY 22 JUNE

Weather fine, wind northwest. Called all hands at 6 A.M. The natives shot one white whale. All hands away at the floe edge fetching four boats in to the land. We got them in all safe, and hauled them on the island.... The cook a good deal better. Night strong wind from the south with heavy rain. All hands nearly drowned out in the tents. We got no sleep all night.

SATURDAY 23 JUNE

...All hands come to the ship. We brought all the boats with us as far as we could get them for the ice [is] breaking up very fast. We have seen a good few fish yesterday and today we are going to send four boats away up to Kingaway....

FRIDAY 27 JUNE

Weather fine, wind north. Called all hands at 4 A.M. Some employed on board of the *Isabel*. We sent the crow's nest up. All the sails loose, drying. Others employed in watering ship. Some of the natives on board. Night fine. Saw several fish. A ship in sight about five miles off, a black barque. She appears to be fishing. She is to the southward of us. Me washing for the steward. We have only got six weeks provisions on board, so we begin to quake for fear.

As THE FAST ICE broke up in Cumberland Sound and the floe edge gradually lost its identity, the whaleboat crews of the *Emma* and *Isabel* began to cruise more extensively from their island base, crossing more than fifty miles of open water to Kingnait Fiord, and sailing a hundred miles up to the head of the gulf, where on 13 July they saw "wite wales and unicorns...thousands thick." The whalemen did not systematically exploit these species, but the time was not far off when the depletion of bowhead whales would cause them to turn their weapons upon the large beluga population of Cumberland Sound, and that of Prince Regent Inlet as well, for alternative supplies of sea-mammal oil. After breaking out of harbour ice on 2 July the steam tender *Isabel* worked in company with the boats, providing food and sleeping accommodation for the crews and standing ready for flensing and the storage of blubber. But the whaling was not very successful. The growing expanse of ice-free water was enabling the whales to disperse more, and the arrival of additional vessels was increasing the hunting competition. In mid-August the ships sailed south to a harbour in Frobisher Bay, from which the boats cruised unsuccessfully for a week. The men then turned to cleaning the ships, touching up the painting, taking on fresh water, and making ready for sea. On 3 September they weighed anchor and started home.

One of the threads running through the journal entries of Albert Whitehouse is the constant interaction between whalemen and Eskimos at the winter harbour. Whaling had opened a narrow window through which many of the native inhabitants of the Arctic could obtain a view of a world beyond their own. It enabled them to meet people who looked different, smelled different, spoke an unfamiliar language, and frequently inserted special clay pipes into their mouths in order to emit smoke from their bodies. The strangers possessed an awe-inspiring technology, which included sailing ships larger by far than the biggest *umiak*, harpoon guns and hurricane lanterns, saws and scissors, mirrors and matches, telescopes and tea kettles, and new materials such as iron, glass, and woven cloth, often in bright colours unknown in the tundra landscape but somehow warming and cheerful. The white men introduced them to new foods, quite different in texture and taste, such as salt beef (in appearance revolting; in smell worse: a people must be totally barbaric to consume the likes of this), porridge (almost the consistency of caribou stomach contents, but without the good taste), hardtack (a good imitation of a slice of driftwood), and sugared tea (delightful!).

The wonderment experienced by the Eskimos during initial encounters with whalemen diminished as contact became more frequent, and it was soon obvious that the strangers were not totally different. They were men, after all, and they seemed intent on hunting *arvik*, the big whale, a fact that struck a responsive chord among the native hunters and excited their admiration. Understanding and communication grew as Eskimos began to participate in whale hunts in return for wages in kind, and as the evolution of a *lingua franca* of pidgin English facilitated the exchange of ideas, the clarification of intentions, and the explanation of procedures. It was true that the foreigners were somewhat authoritative and assertive at times, and inclined to take far too much for granted in the territory of Inuit, but the natives gazed with envy upon their weapons, implements, and materials, some of which promised to make life in the Arctic less difficult and more secure, and they reasoned that friendly co-operation might be the most appropriate response to the intrusion of the whites.

In Cumberland Sound economic ties had developed quickly between whalemen and Eskimos after the first contact in 1840. It was the captains who took the initiative, hiring a few individuals to fill gaps in boat crews at first, and later outfitting boats with all-native crews to work on behalf of the vessels. The Eskimos delivered the blubber and bone of any whales they killed and were paid off in material goods at the end of the season. By the time of the *Emma*'s 1859 voyage this practice was well established. Approximately 350 Eskimos were said to be living around the Sound and with a dozen or so ships cruising there in summer Eskimos men had plenty of opportunities to participate in the whaling and thereby obtain some of the manufactured articles they coveted. There was positive feedback in the system: as they enriched their material culture with boats, harpoons, lines, lances, flensing tools, and guns, obtained through employment, and as they gained experience in European methods of whaling, they found it easier to secure summer employment with the whalers; and through regular employment they continued to improve their whaling outfits.

The practice of wintering greatly reinforced these economic ties, so that the most enterprising Eskimo men and their whaleboat crews could work almost continuously through the seasons, year after year.

Working for the wintering ships did not mean giving up all the traditional activities of life on the land. There were sacrifices to be sure, but in general a ship's native could continue to hunt and travel as before. The difference was that he was now expected to secure an amount of meat much greater than his needs in order to feed the whalemen. The ship captain reduced the hunter's own requirements by providing some shipboard food to him and his family, and contributed to his hunting effectiveness by outfitting him with a gun and plenty of ammunition. Although the hunter would travel far afield at times during the autumn and winter, the nomadism of his family was reduced; they tended to remain at the whaling harbour, where the captain handed out daily bread, coffee, and sometimes molasses. This centralization of Eskimo population was one of the most important consequences of whaling. (Something similar occurred a century or so later, when the Canadian Government encouraged people to abandon their small, scattered camps and amalgamate in a few large administrative centres where they could conveniently receive education, health care, and other benefits.)

Life at the winter harbour, with so many new faces and such a variety of fascinating activities, must have had a certain glamour and attraction to the Eskimos, like the lure of bright city lights to rural dwellers. It had its drawbacks as well. Concentrations of Eskimos were highly susceptible to alien diseases. Something as apparently harmless as influenza or measles could sweep through a native settlement with devastating results. The whalemen, like other sailors the world over, liked to have "a girl in every port," and places such as Kekerten and Niantelik were no exception. But one consequence of licentious behaviour was venereal disease, including gonorrhea and syphilis, and these afflictions were soon present among the Eskimo population. Another product of friendly relationships with Eskimo women was, of course, children, sired by sailors who had not the slightest intention of assuming any responsibility for their offspring, and who frequently had left the country by the time the children were born. It is unlikely that many of the children remained fatherless, however; the flexibility and tolerance embodied in Eskimo adoption attitudes and practices assured them of being raised within some family unit without discrimination.

A number of whaling captains kept the welfare of the Eskimo population in mind and were perceptive enough to recognize that if unruly crews behaved as they wished in native communities they could inflict great and lasting harm. In 1848 Captain Parker had petitioned Queen Victoria to assist destitute Eskimos of the whaling regions, and he took a man and a woman from Cumberland Sound back to England to support his request. In the 1850s Captain William Penny had made repeated attempts to have a permanent mission established on Baffin Island, to counteract the pernicious influence of whaling crews. But these gallant efforts came to nothing, and while responsible captains sailed to Davis Strait without liquor on board, held regular services of worship during the voyage, kept a tight rein over the conduct of the men, and exercised strict control over their behaviour among the Eskimos, their good works were often negated by irresponsible and ineffective masters, such as Simpson of the *Emma*, who stocked their ships with liquor and let the men do just about anything they wanted.

In the summer of 1860, as the *Emma* emerged from winter quarters in Cumberland Sound, a number of British and American ships were arriving for the whaling season. At least eight vessels, seven of them American, had been outfitted for a winter in the Arctic. Over the last ten years the cumulative experience of several masters and many old hands, and the co-operation of the Eskimos, had effectively removed most of the fear of wintering. A few of the veterans had even come to perceive the Arctic as the "friendly" place that the explorer and writer Vilhjalmur Stefansson was to popularise a half-century later. This optimistic perception, however, was seldom shared by men undertaking their first arctic voyage, to whom the unfamiliar environment often appeared hostile. From time to time some men, apprehensive about the daunting prospect of spending a winter in the Arctic, and discontent with conditions on board their vessels, chose to desert ship and travel south by whatever means they could find. But as the next chapter shows, such desperate adventures had little chance of success.

Open whaleboats, although seaworthy, were not ideal for long voyages. (Public Archives Canada, C-46970)

Eleven
Desperate Voyage

Larboard boat missing. A boatsteerer and six men have deserted, taking clothes, cook's provisions, boat's gear.
Logbook of the *Ansel Gibbs*, 5 August 1860

DESERTION OF SEAMEN from their ships was a common occurrence in the days of sail, and it caused serious problems for shipowners, masters, consular officials in foreign ports, and many others. Conditions of shipboard life in the world's navies, mercantile marines, passenger services, and whaling fleets were generally rigorous, uncomfortable, and unhealthy, and voyages were often of long duration, especially before the completion of the great short-cuts of the Suez and Panama canals in 1869 and 1914. Whaling vessels in particular spent many months, often several years, at sea, cruising until their holds were full of oil, and frequently remaining out even longer if they could manage to send produce home from some distant port on board another vessel. Many American and European whalers circled the earth during their voyages, crossing the Atlantic, Indian, and Pacific oceans before returning home three or four years later, and when they finally arrived it was rarely with the same crew that had begun the voyage. On American whaleships only one-third of the original crew usually remained by the end of a voyage. The American whaler *Addison*, for example, started in 1856 from New Bedford, Massachusetts, with 33 men, sailed around Cape Horn to Hawaii, cruised in the Pacific, went twice to Bering Sea and Strait, reached New Zealand, and returned home by way of the Horn, arriving in 1860 with only ten of her original crew left. Altogether, another sixty-nine men had been signed on along the way during the four-year voyage. Whaling crews were whittled down by discharges, arrests and detention on shore, hospitalization following accident or illness, deaths, transfers to other ships, and not least of all by desertions. Twenty-two seamen deserted from the *Adeline* during a three-year voyage (1866–69); nineteen fled from the *Isaac Howland* (1854–59); the *Montreal*, during a five-year cruise (1857–62), lost thirty hands.

Dissatisfaction with life on board, combined with the attractions of a number of enchanting islands and ports along the route, induced sailors to leave their vessels clandestinely and follow their inclination ashore. Many of the experienced professional whalemen might be willing to tolerate crowded and unsanitary quarters, bad food, scurvy, harsh treatment, and long periods of boredom punctuated now and then by hard work and danger, but the farm boys and city slickers who went to sea on a whim or in search of adventure, and who stood to earn little if anything from the voyage, were less tolerant of such conditions and more susceptible to the seductive atmosphere of foreign places. Men jumped ship in the Marquesas and Moluccas, in Samoa and the Sandwich Islands, in Mauritius and Manila—wherever there was the right mix of shipboard discontent and shore attractions, and where an escape could be managed. Some deserters were apprehended by ships' crews or by the authorities ashore. Others evaded capture for a while but in the end shipped again on whaling vessels. But many rejected the whaling life forever and a significant number took local wives and settled down in remote corners of the world, where their social and economic impact on native communities was sometimes considerable.

While conditions on board the arctic whalers often provided substantial grounds for desertion, a man thought twice before deserting ship on the barren arctic coasts, where survival demanded sophisticated techniques of providing shelter, securing food, and generating heat. Nevertheless, there were men who either calculated or ignored the risks and fled ashore. A few of them married Eskimo women and remained in the Arctic, "going native" as their shipmates termed it. But most of the deserters headed south for civilization, craving warmth and greenery, family and friends. How to get there was the problem, and in their ignorance of fundamental geographical facts many underestimated the hardships involved and attempted foolhardy voyages that had little chance of succeeding.

In the year 1895 (more than 30 years after the events that are the focus of this chapter), four men deserted from the American whaler *Canton*, frozen into winter quarters at Cape Fullerton, on the northwest coast of Hudson Bay. They started to *walk* out of the Arctic in the middle of December intending to reach the United States—more than a thousand miles to the south—by way of Hudson's Bay Company posts at Churchill and Fort Nelson. The men were soon overtaken and escorted back to the vessel in a blizzard, during which one of them perished.

At about the same time, on the far western margin of the Canadian Arctic, American whalemen at the wintering harbour of Herschel Island were responding to the lure of Klondike gold. In March 1896 a dozen or more men dreaming of quick fortunes deserted from the whaleships and attempted to reach the goldfields six hundred miles south by dog sled. Within a few days ship's officers had overtaken them, killed one man and captured six.

Deserters' dreams nourished foolish schemes. Rarely, if ever, was flight undertaken with adequate food and equipment because preparations had to be carried out in secret and equipment obtained by theft. Away from the ship, confidence usually faded quickly as the men encountered the harsh reality of cold, hunger, and fatigue. But most were either too proud or too stupid to turn back.

In Cumberland Sound, in August 1860, on board two of the American whalers intending to spend the winter, dissatisfaction with shipboard conditions and the gloomy spectre of ten months frozen into harbour ice thousands of miles from home generated thoughts of desertion. The idea flitted secretly through individual minds, gained feasibility in hesitant, illicit discussions, and eventually evolved into a secret plan of escape shared among nine men. Seven comprised an entire boat's crew on the Fairhaven ship *Ansel Gibbs* and on the night of 4 August, while they held the deck watch, they quietly lowered a boat, rowed to the New Bedford ship *Daniel Webster*, picked up two friends, and made sail silently for civilization.

The deserters intended to proceed south out of Cumberland Sound, cross the mouth of Frobisher Bay, pass Resolution Island, sail westward through Hudson Strait, cross Hudson Bay to the Hudson's Bay Company post at York Factory, and then continue overland. For this ambitious voyage in an open boat not thirty feet long, they carried some whaling gear, a compass, two guns, some ammunition, a small cooking stove, twenty pounds of bread, a few blankets, and some clothes—woefully inadequate stores for such an adventure. As to the experience of the men, seven were green hands on their first whaling voyage and the other two, although they had shipped on whaling voyages before, were novices in the Arctic. With confidence and rashness born of ignorance they all set out to sail 1500 miles through northern waters, with no chart and no sextant.

Whaleboats under sail. (General Synod Archives, Anglican Church of Canada)

Three days after leaving their ships the men had reached a point between the mouths of Cumberland Sound and Frobisher Bay. They were already running low on food, but by good fortune they met up with the whaler *George Henry* of New London, Connecticut. Captain Sydney Buddington, who a decade before had been in charge of the first group of whalemen to intentionally winter ashore (chapter 12), provided some assistance, and after he demonstrated the futility of attempting the Hudson Bay voyage the men decided instead to cross Hudson Strait and work southward along the coast of Labrador until they met up with fishing vessels or native settlements. It was a sensible and fortunate decision, to which most of them probably owed their lives.

The chance encounter with the deserters was witnessed by Charles Francis Hall, an American heading north aboard the *George Henry* on his first attempt to discover if any survivors of Sir John Franklin's lost expedition were still alive. Later, in his published narrative, Hall included the following firsthand account by one of the deserters, which had been sworn before a Newfoundland magistrate. In it John Sullivan revealed the shocking outcome of the men's desperate hunger during the remainder of the boat journey.

My name is John F. Sullivan. I left my home in South Hadley Falls, Massachusetts, about the first of March, 1860, for Boston. I remained in Boston until the twentieth of the same month. I applied at different offices for a chance to ship; being a stranger in the place, and a green hand, I found it very difficult to get a berth to suit me. At last I got a little discouraged, and that day signed my name at No. 172 Commercial Street, Boston, and left for New Bedford, Massachusetts. Next morning, I shipped to go aboard of the ship *Daniel Webster*, then laying at New Bedford, but to sail the same day on a whaling cruise to Davis Straits, to be gone eighteen months.

I left New Bedford in the *Daniel Webster* on 21 March 1860. There were forty of us in the crew, all told. We had very rough weather for many days after leaving, which caused many of us to be seasick; I suffered from it about three weeks; after that time I began to recruit. There was nothing happened of any consequence worth mentioning until we passed Cape Farewell, about the last of May. After that we had quite a hard time, working the ship through the ice; occasionally, however, we made out to get her through, and came to anchor 6 July 1860 [probably in Kikastan Islands].

We spoke many vessels going in. I will name some of them: the *Hannibal*, of New London; the *Black Eagle* and *Antelope*, of New Bedford; the *Ansel Gibbs*, of Fairhaven; the *Pioneer*, of New London. These vessels were anchored very close to one another in the harbour; the crews were at liberty sometimes to pay visits to each other; each one would tell how he was treated, several complained of very bad treatment, especially the crew of the *Ansel Gibbs*; they were planning some way of running away for a long time, but they found no opportunity till the fourth of August.

My shipmate, whose name was Warren Dutton, was aboard that day, and heard a little of the conversation, and he joined in with them, and said he would go, and perhaps one or two more of his crew. He immediately came aboard and informed me; and he pictured everything out so nice, that I finally consented to go with him. We had no great reason for leaving our vessel; we could not complain of very bad treatment aboard; all we could complain of was that we were very badly fitted out for such a climate; and, after we arrived there, hearing of so many men that died there last winter of scurvy, we were afraid to remain there, for fear that we might get it. We thought that by running away, also, we would be all right, but we were sadly mistaken.

After it was agreed upon to leave, each one was busy making preparation for a start. I, with my shipmate, packed what few things we thought would be necessary into a travelling-bag which belonged to me; we then crept into the hold, and filled a small bag and a pair of drawers with hard bread, and waited for an opportunity to hide it on deck, unknown to the watch. After we succeeded in

that, we made a signal to the other crew that we were ready. It being boats' crew watches aboard the *Ansel Gibbs*, they every one of them left; they found no difficulty in lowering away the boat, which after they did so they lowered themselves easily into her, and soon paddled under our bows; we then dropped our traps into her, and, taking with us two guns and a little ammunition, got into her, and soon pulled around a small point out of sight of the vessels. The names of the crew that left the *Ansel Gibbs* are as follows: John Giles, boatsteerer, John Martin, Hiram J. Davis, Williard Hawkins, Thomas Colwell, Joseph Fisher and Samuel J. Fisher.

At eleven o'clock at night, on the fourth of August, we left the vessels in Cumberland Straits [Cumberland Sound], latitude 65°59′, about five miles from Penny's Harbour. Although it being a little foggy, with a fair wind we stood across the Straits. When about halfway across we dumped overboard a tub of towline to lighten the boat some. We had nothing but a small boat-compass to guide us; we had no opportunity of getting a chart before we left, and not much of anything else.

We made the other side of the Straits by morning; then, by taking the spy-glass, we thought we could perceive a sail in chase of us, but we soon lost sight of her. The other crew were depending mostly on us for bread, as my shipmate informed them that we had a better chance to get it out of the hold; their bread lay close to the cabin; so, what bread they had, with ours, would not exceed more than twenty pounds. We all saw that the bread would not last long, so each one desired to be put on allowance of one biscuit a day to each man. We hoped, by the time that was gone, to reach some place where we could find help. We made a very good run the first three days, sleeping at night in the boat; on the fourth day we fell in with the barque *George Henry*, Captain Budington, of New London. He asked us aboard; the boatsteerer acted as spokesman. The captain told us we were very foolish to leave the vessels to undertake so long a trip. I believe he would have taken us

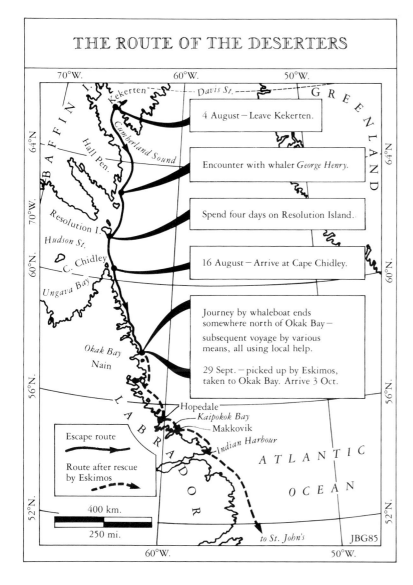

THE ROUTE OF THE DESERTERS

4 August – Leave Kekerten.

Encounter with whaler *George Henry*.

Spend four days on Resolution Island.

16 August – Arrive at Cape Chidley.

Journey by whaleboat ends somewhere north of Okak Bay – subsequent voyage by various means, all using local help.

29 Sept. – picked up by Eskimos, taken to Okak Bay. Arrive 3 Oct.

Escape route

Route after rescue by Eskimos

400 km.

250 mi.

That night we made a 'lee', found some moss, and made a fire; before we ran in we shot a small duck, which made a good stew for all hands. Two days after this we shot a white bear; he was in the water when we shot him, and there being a heavy sea on at the time, we could get no more than his hind quarters in; them we skinned – the rest we could not save. That night we managed between us to cook it, as we were divided into watches, two in each watch; by doing so, we could watch the boat and keep her with the tide. We kept on in this way, always tracking the shore, and at night going ashore to lay on the rocks, with our boat's sail over us for shelter.

We had very rough weather in crossing the Straits [Hudson Strait]. We were on Resolution Island four days, waiting for a fair wind; we got it at last, but so strong that it came very near swamping our little boat many times through the night. It kept two of us bailing water out all the time, and we were glad to reach the land, after being in the boat thirty hours, wet to the skin. What bear's meat and bread we had was most gone by this time; there was nothing left but a few crumbs in the bottom of the bag. There was nine parts made of the crumbs; then they were caked off, each man taking his share.

On the sixteenth of August we made Cape Chidley; on the twentieth we divided the last crumbs; after that we picked up what we could find to eat. We found a few berries and mushrooms; we suffered very much from the cold, very seldom having a dry rag upon us.

We continued on in this condition until the third of September, when, to add to our misfortune, Williard Hawkins and Hiram J. Davis (who we called 'the doctor') ran away from us that night, and took with them everything that was of any use to us; they even took the boat's compass, and left us in a miserable condition, with our boat broadside on the beach. It being their watch, they made out to get off. We thought it was useless to make chase after them, so we let them go. It then commenced to rain, and there was heavy sea rolling in, and, weak as we

all if we wished to stay; but as we had left a whaler, we did not like to go on board another, as he was also going to remain there through the winter; so we were determined to push along, as we had been foolish enough to start in the first place. However, before we left, he gave us a small bag of bread, a piece of salt pork and some ammunition; also a chart. We then bade him goodbye, and set off again.

were, we found some difficulty in shoving the boat off. However, after a hard tug, we succeeded, and then pulled out some ways; we then up sail; it was not up long before it blew so strong that it carried away the mast. We then ran in under a jib, and made a lee. About half an hour after we landed my shipmate died of starvation. The evening he died, Samuel Fisher proposed to eat him; he took his knife and cut a piece off the thigh, and held it over the fire until it was cooked. Then, next morning, each one followed his example; after that the meat was taken off the bones, and each man took a share. We stopped here three days. We then made a start; but the wind being ahead, we were obliged to put back. Here we stopped two more days. During that time the bones were broken up small, and boiled in a pot or kettle we had; also the skull was broken open, the brains taken out, and cooked. We then got a fair wind but as we got around a point, we had the wind very fresh off shore; we could hardly manage the boat; at last we drove on to an island some ways out to sea; we got the boat under lee of it; but the same night we had a large hole stove into her. Being unable to haul her up, we were obliged to remain here eight days: it was on this island they tried to murder me.

The third day we stopped here, I was out as usual picking berries, or anything I could find to eat. Coming in, I chanced to pick up a mushroom. I brought it in with me; also an armful of wood to keep. While kneeling down to cook the mushroom, I received a heavy blow of a club from Joseph Fisher, and before I could get to my feet I got three more blows. I then managed to get to my feet, when Samuel Fisher got hold of my right arm; then Joseph Fisher struck me three more blows on the arm. I somehow got away from them, and, being half crazy, I did not know what to do. They made for me again; I kept begging of them, for God's sake, to spare my life, but they would not listen to my cries. They said they wanted some meat, and were bound to kill me. I had nothing I could defend myself with but a small knife; this I held in my hand until they approached me. Samuel Fisher was the first to come toward me; he had a large dirk-knife in his hand; his cousin was coming from another direction with a club and a stone. Samuel came on and grasped me by the shoulder, and had his knife raised to stab me. I then raised my knife, and stabbed him in the throat; he immediately fell, and I then made a step for Joe; he dropped his club, and then went where the rest was. I then stooped down to see if Samuel was dead; he was still alive. I did not know what to do. At this time I began to cry; after a little while the rest told me to come up; they would see there was nothing more done to me. I received four deep cuts on the head; one of the fellows dressed them for me, and washed the blood off my face. Next day Samuel Fisher died; his cousin was the first one to cut him up; his body was used up the same as my unfortunate shipmate's.

After a while we managed to repair the boat, and left this island. We ran in where we thought was main land, but it proved to be an island. Here we left the boat, and proceeded on foot, walking about one mile a day; at last we reached the other side of the island in four days; they put back again to the boat. It took us four days to get back again. When we got there, we found the boat was stove very bad since we left her. We tried to get around the island in her, but she sunk when we got into her; we then left her, and went back again to the other side of the island, to remain there until we would die or be picked up. We ate our boots, belts, and sheaths, and a number of bear-skin and seal-skin articles we had with us. To add to our misery, it commenced to rain, and kept up for three days; it then began to snow. In this miserable condition we were picked up by a boat's crew of Eskimo on the twenty-ninth of September, and brought to Okoke [Okak] on the third of October. The missionaries did all that lay in their power to help us along, and provided us with food and clothing, then sent us on to Nain, where we met 'the doctor,' who was picked up three days before we were. He reported that his companion died, and told many false stories after he was picked up.

The missionaries of Nain helped us on to Hopedale; from there we were sent on to Kibokok [Kaipokok Bay], where two of us remained through the winter. One stopped with a planter, named John Lane, between Nain and Hopedale; the doctor stopped with John Walker until March, when he left for Indian Harbour; the remaining two, Joseph Fisher and Thomas Colwell, also stopped with planters around Indian Harbour. Mr Bell, the agent at Kibokok, kept two of us until we could find an opportunity of leaving the coast. We left his place about the tenth of July, and came to Macovie [Makkovik], waiting a chance to get off.

Captain Duntan has been kind enough to give me a passage; my companion was taken by Captain Hamilton, of the *Wild Rover*. We have had a very pleasant passage so far, and I hope it will continue so.

Sir, I hope you may make it out; it is very poor writing, and was written in haste. John F. Sullivan.

FOR THE MEN who had felt unable to tolerate harsh treatment and insufficient food aboard their vessels the awful experiences of the desperate boat voyage to Labrador had proved immeasurably worse. Three had lost their lives (one, at least, by murder) and the rest had survived only by the slimmest margin, after resorting to the final degradation of cannibalism. They owed their salvation, not to their own survival skill or to any spirit of mutual co-operation—for they had given no evidence of either—but to the kindness and efforts of the Eskimos, Moravian missionaries, and others, who rescued them, took them in, fed them, clothed them, and transported them south in stages to Newfoundland. They owed their lives to these people of Labrador, and yet they made nuisances of themselves wherever they stayed. The Moravians of Okak later reported that the deserters "...were thoroughly coarse, ungrateful beings. There was something mysterious about them, and perhaps the less is said, the better. We were truly thankful that some of our people were found willing, so late in the season, to take them...to Nain."

Yet there was a positive dimension to this tragic escapade. It appears to have made a strong impression upon Charles Francis Hall, then at the outset of his extraordinary career in arctic exploration. During two subsequent winters based on the whaler *George Henry* in Cyrus Field Bay, Hall sledded south into Frobisher Bay, charted its coastline, disproved the idea of its opening westward into Hudson Strait or Foxe Basin, discovered a number of relics of Frobisher's expeditions three centuries before, and observed and described Eskimo life. His second expedition followed closely on the first and lasted five years, during which he travelled west from Hudson Bay to King William Island, discovering graves and other relics of the Franklin expedition. Finally, in the *Polaris* expedition of 1871–73 he sought the Northern Hemisphere's ultimate prize, the North Pole—an unsuccessful venture in the course of which Hall himself was murdered.

Hall was an innovator in arctic exploration, adopting Eskimo techniques of travel and subsistence in an age when the customary approach of most non-natives was tradition-bound and inappropriate. By utilizing techniques that the unfortunate men of the Franklin expedition had neglected to adopt Hall was able to reach the scene of their disaster and partly reconstruct the story of their end. To what extent were Hall's determination and effectiveness inspired by his brief encounter in 1860 with the deserters from the *Ansel Gibbs* and *Daniel Webster*? Shortly after the event he wrote:

If these nine men can undertake such a voyage, and under such circumstances, with so little preparation, why should not I, having far better means, be able to accomplish mine?

George Tyson. (National Archives and Records Administration, Washington, D.C.)

Twelve

Reflections of a Whaling Captain

I ... returned to my old business in the manufacture of ironware, but very soon grew tired of it, and again longed for the sea. GEORGE TYSON, 1874

I F THE DISCOMFORT, DEPRIVATION, AND DANGER of the arctic whaling routine drove some men to desertion, the same ruggedness, simplicity, and adventure of life on the arctic seas attracted many others into a succession of voyages or even a career in whaling. One Edward Moore, apparently of Hull, sailed every year but two from 1808 to 1821 to the Greenland or Davis Strait whaling grounds. Another man, Thomas Shakesby, made a dozen arctic whaling voyages between the years 1801 and 1821. Alexander Sandison, a Shetlandman, sailed twenty-seven times to Davis Strait. The Whitby whaleman William Scoresby, Jr., who made an enormous contribution to scientific knowledge of the arctic regions, completed more than twenty Greenland voyages before turning to a career in the church at the age of thirty-three. William Barron of Hull made seventeen voyages, most of them into Davis Strait, between 1849 and 1865, serving in every position from apprentice to master. One of the champions of long service was Captain David Gray of Peterhead, who completed no fewer than forty-nine voyages before retiring in 1893.

George Tyson, whose published memoirs are excerpted below, was another whose best manhood years were taken up primarily by arctic voyages. He was attracted into the whaling trade by dreams of polar exploration, rose to become master, wintered in the Arctic on eight occasions, and participated in a number of discovery expeditions.

During the seafaring careers of men such as these the Arctic generally claimed more time than their homes and families, and inevitably called them away during the most pleasant time of the year. The joys of summer were sorely missed. The captain of the *Thomas* (chapter 4) complained in 1834 that during twenty-one years of arctic cruising he had been unable to taste fresh strawberries. Captain George Tyson bemoaned the fact that for two decades he had never been able to enjoy an American Independence Day celebration at home. And William Barron in the twilight of his life, admitted with regret, "I never saw either blossom or fruit upon the trees, and my eyes and senses were never blessed with the scent of growing flowers, the sight of ripening corn, or the subsequent harvest operations." The whaling life, he said, was "frequently one of great privation, and at all times of deep denial and not a little danger." Yet the beauty, mystery, and unpredictable violence of the arctic seas, and the savage thrill of pursuing the great whales, lured men back again and again.

Because they regularly confronted the powerful antagonists of sea, ice, and rock with flimsy wooden hulls and canvas sails as a means of livelihood, arctic whaling masters were quite different from the leaders of exploring expeditions, whose reputations were sometimes based on a single voyage, or two or three. And whereas the exploits of the whalemen were apt to be well-known in their home ports they received far less national publicity than the achievements of explorers. There were whaling masters who could hold up their heads in any company for qualities of enterprise, determination, skill, and leadership, but their names (unlike those of Parry, Peary, Ross, and Hall) were never engraved on the public consciousness. They were unsung heroes. Collectively, whaling masters and crews accomplished much more than the mere delivery of oil and baleen to European and American ports. They discovered new lands, noted unfamiliar animals, plants, and ice features, and extended European culture to the native people of the arctic regions. Furthermore, their activities in northern no-man's lands and seas led in the nineteenth century to the formal expression of British territorial aspirations in the Arctic, and the subsequent acquisition by Canada of the vast archipelago lying between Greenland and Alaska.

British vessels dominated the Davis Strait whale fishery during the nineteenth century. American voyages from 1800 to 1911 numbered approximately 175, only six percent of the total number of whaling voyages in the period. But in one respect – the initiation and practice of wintering – the contribution of American whalemen was considerable. It was part of the crew of the New London ship *McLellan* that carried out the first intentional wintering in 1851–52, and in the next half-century American vessels comprised almost half of the hundred-odd winterings in the region, despite the fact that British ships were far more numerous on the whaling grounds. The peak decade was 1860–70, in which two-thirds of all the whalers wintering in the Davis Strait region were American.

In the autobiographical sketch below George Tyson recounts in detail his first whaling voyage, including the pioneer wintering experiment of the *McLellan*'s men, an event that was to lead to a more intensive relationship between whalemen and Eskimos in Cumberland Sound (chapter 10) and to a system of year-round shore whaling and trading stations (described in chapter 15). He also tells of a curious encounter with an abandoned ship among the ice floes of Davis Strait in 1855.

George Tyson "in his Arctic costume." (National Archives and Records Administration, Washington, D.C.)

I was born in the State of New Jersey [in 1829]; but in early infancy my parents removed to New York city, where I received my early education, and when of suitable age I commenced work in an iron-foundry—my parents, like nearly all others, desiring to keep their sons upon the land. But my heart was always on the seas, and particularly I longed to see something of the arctic world; the names of Ross and Parry and Franklin had seized upon my imagination, and I longed to follow in their track. To witness the novel scenes, and to share in the dangers of arctic travel, was at that period the height of my ambition; and while watching fiery liquid ore that was presently to appear in the shape of grates and fenders, my fancy was off among the icebergs; and, despite the dicta of Shakespeare, I sometimes almost managed to cool my heated brow with "thinking of the frosty Caucasus," I was disgusted with shop labour; and as no opportunity offered for joining any arctic exploring party, I concluded to do the next best thing, and ship in a whaler, which at least would bear me a few degrees toward the coveted regions of perpetual ice.

In execution of this intention, I shipped on board a New London whaler, the bark *McLellan*, Captain William Quayle, in 1850 [1851, in fact] when I was about twenty-one years of age. The *McLellan* was bound to Greenland and adjacent seas. It was Captain Quayle's intention, however, to take the sealing on the coast of Labrador first, and for that purpose sailed very early in the season, leaving New London on the seventh of February, 1850 [1851].

After being out a few weeks, one of our shipmates died, and was buried at sea. This seems a slight thing to record, because the poor man has left no historic name behind him but no one who has not experienced it can realize the great solemnity of a burial at sea, especially when witnessed for the first time by a young person unhardened to the vicissitudes of life. When we deposit our friends in the ground, there seems something left of them; we can at least visit their graves, and adorn them with flowers, and fancy that they know we still care for them; but when the poor discarded body slides over the ship's side, and strikes the water with that heart-sickening *thud*, it appears as though we were giving up our late friend to a more certain and eternal separation. The imagination follows it, indeed, for a while, along the known currents which set to or from the ship, but beyond that we know not its journey, or whither it is carried—whether it ever comes to rest, or is ceaselessly borne about by the ever-shifting waters, until the continual friction first denudes the body of its covering, and then the bones of its flesh, or perhaps, that it is destined to furnish a ghastly meal to some monster of the deep.... Nevertheless we know that at the last they shall not be forgotten. We have the promise that "the sea shall give up its dead."...

[We arrived]... off Resolution Island in the early part of March; and I shall never forget the terrible weather we experienced in this vicinity.... I have been much farther north since then, but I never remember feeling the cold more, except while on the ice floe; and I don't think I should then, only for lack of suitable clothing and sufficient food.

We were now compelled, through stress of weather, to run into the pack of ice for protection. There we found the wind less violent, and no sea—the unusually heavy floe entirely destroying the force of the huge waves as they beat against it. In the ice here we found many seals of the "bladder-nose" [hooded] species—so called from the bladder, or hood, which they have on their head; this hood, when they are excited or angry, they can expand to a great size, and then you may know they are ready to defend themselves. It was from this early experience off the west coast of Greenland that I derived my knowledge of the haunts of this kind of seals; and it was this which gave me hope through that dismal arctic winter, and enabled me to hold out encouragement to the rest of the party while we were on our long drift, knowing that once we could reach this latitude, all fear of starvation would be at an end.

On this my first voyage we killed numbers of these seals,

and many a good battle we had with them. The male, in particular, will defend his family to the death. These seals are quite large, and are taken, like the whale, mainly for the oil.

Finding no whale off Resolution Island, we next bore away for the coast of Greenland, it being the captain's intention to work his vessel...to Pond Bay where we were sure to find whales.

We first sighted the land at Holsteinborg, a settlement in the south of Greenland. The land in that vicinity, and, as I afterward discovered, nearly the whole coast, is high and mountainous, presenting a most desolate appearance. If I had not known the fact, I could scarcely have believed that it contained any inhabitants. But we soon had ocular demonstration of this; for, getting the ship in the ice, we were almost immediately "beset," and very shortly after a number of these hardy sons of the North were seen coming over the broken ice to pay us a visit; and their first appearance convinced me how much more one is impressed by *seeing* than by reading or hearing. Of course I had always heard of the small stature of the Eskimo, and thought I knew just how they looked; but when I saw these little creatures approaching—the men less than five feet, and the women not more than four—I realized the difference of race in a way one can not do without seeing. I have often thought since that nature made them so small that they might travel the more easily over the thin ice and the snow, as they often have to do, in pursuit of seals and other game. If they had been made as large and heavy as many of the white race, they would be far worse adapted for the mode of life which that barren country forces them to adopt. As many of them were dressed in sealskins, with round sealskin caps on their heads—and, when laid horizontal on the ice, about the length of the smaller kind of seals—I could not help thinking but that God had made them thus, with their brown faces, so that they could imitate the creatures, and so decoy and catch them, which they often do. Holsteinborg, I afterward found, consisted of about a dozen huts, or houses, and less than fifty inhabitants.

The ice did not detain us long here. It soon opened, and we proceeded northward to Disko, where we were again stopped by the ice. Disko is a regular rendezvous of the English whaling fleet, as well as being frequently visited by the American whalers....There are over twenty houses here, and, I was told, seventy or eighty people. I have since ascertained that the Danes who come out here in the governor's suite, and others who visit the country for commercial purposes, and stop any length of time— especially those who intend to make it their home—not frequently marry native women; so that at some of the settlements you may see a family where the children have the light, flaxen hair of the Dane, and the dark, bronzed cheek of the native. This mixture makes a curious physiognomy.

The highest point to which we sailed on my first voyage was called the "Devil's Thumb," in Melville Bay. This "thumb" is a large pointed rock, like an immense Bunker Hill Monument, that rises perpendicularly to the height of five or six hundred feet; and as it stands on a very high, rocky island, its topmost point is probably fifteen or sixteen hundred feet above the level of the sea. There is a great deal of superstition about this 'Devil Thumb'—partly, I suppose from its name; but to those who know the difficulty of steering among the icebergs which abound here, and the cross-currents which swirl the bergs about, and of course treat ships the same, there is no need to go beyond nature for objects of terror.

...When off this point we were again beset in the ice, and...as July was now on us, Captain Quayle concluded to turn south before crossing to the west, so as to take the whales as they came from the north, which they commence to do in August.

When near the Duck Islands, we saw two strange vessels, with colours set; found them to be Americans; and, on speaking, learned that they were the two vessels sent out by the government to aid in the search for the brave and lamented Franklin, under the command of Lieutenant De Haven, a brave and energetic naval officer. They had been drifting in the pack all winter, and were now slowly working their way north again. Captain Quayle supplied them with potatoes and whatever else he could spare. De Haven returned home that fall.

After parting with the explorers, we took the ice off Disko about lat. 69°14′N, long 53°30′W, and endeavoured to get west by taking advantage of every opening in the ice, and soon after sighted the west coast of Davis Strait. Then a thick fog set in on us. At this time we were surrounded by whales; but it is almost impossible to take them when the ice is loose and broken, on account of their running under the large heavy floes to escape, taking the line with them. But we tried our luck, and fortunately captured two.

... All the large whales of this region are 'baleeners'; that is, the mouth and upper jaw are furnished with the baleen, or whalebone, of commerce. When a whale is fastened to the ship, and the cutting and stripping of the blubber is going on, the head is usually first severed from the body for convenience in getting at the baleen; but a boat *can* enter the mouth of the whale, and, if necessary, several men could at the same time stand upright and be at work, removing the whalebone from the upper jaw, the head of the whale being about one-third of the bulk of the creature. The whales change their haunts frequently. When they are too closely followed in one sea, they go to other grounds. In 1810, and a little later, whales were plenty in Baffin Bay and along the west coast of Greenland; then, being too sharply followed, they migrated to Hudson Bay; and when they were followed there they became, after awhile, very scarce, and the next we heard that they were very plenty off Bering Strait, on the other side of the continent. Is it possible that the whales have the only practicable Northwest Passage to themselves? I remember that brave old whaler Scoresby tells in his book about several whales being found in Bering Strait with harpoons in them bearing the mark

and date of ships which sailed only in Baffin Bay; and later sailors' yarns have revived such stories, which I always doubted. But Professor Maury, in his *Geography of the Sea*, puts it down for fact without any question. If the whales do have an "underground railroad" to the Pacific, they undoubtedly come and go in both directions; for not many years after they were reported plenty in Bering Strait they were back again to the northwest part of Hudson Bay and Davis Strait. But certain it is, that no whale has ever been found on this side of the continent bearing any evidence that it had travelled from the Pacific. I think the scarcity and plenty, and, within certain limits, the changing of their haunts, is explained by the fact that when they become unprofitably scarce in one location the whalers go to another; and thus give them, for two or three seasons, a chance to breed again....

I once had, when I was boatsteerer, quite an adventure with a whale which was determined not to die. It was a large and valuable baleener. Soon after the boat was lowered we got alongside. As I rose to heave the harpoon, it seemed, almost in an instant, that the whale had plunged down to the bottom of the bay; as the rope uncoiled and went over the gunwale it fairly smoked with the intense rapidity of the friction, and I had to order it "doused" to prevent its taking fire. It came, too, within a hair-breadth of capsizing us. Fortunately the line was over seventy fathoms long, and of the strongest kind. After she plunged we followed on, it taking all our strength to bring the boat near enough to her to keep the line slack. She stayed underwater the first time so long that we thought she was dead and sunk. It was nearly an hour before she rose; and when she did, the jerk almost snapped our strong line, already weakened by the friction and unusual tension.

As soon as she appeared she began to beat the water with her flukes, and swirled around so that it appeared impossible to get a lance in her; and, while I was endeavouring to do this, our line parted, and away she went, carrying the harpoon with her. We followed with all the speed we could

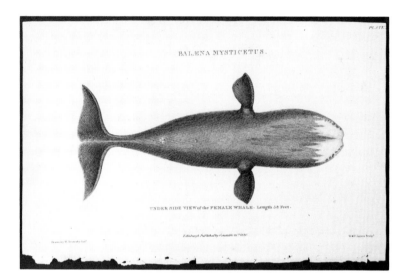

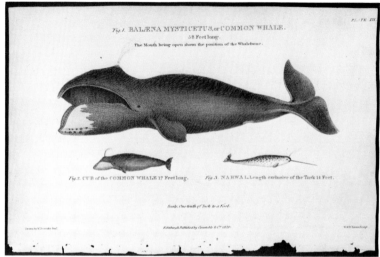

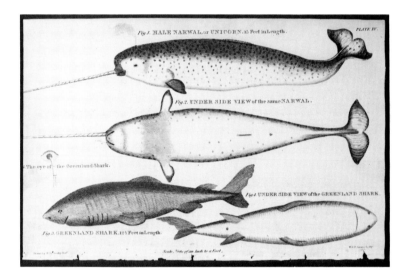

force, and, at last, after several hours' hard pull, came up with her. She seemed to know we were following, and several times disappeared, and then coming up to blow, perhaps half a mile off; but we were bound to have her. On and on she went, on and on we followed. The moon was shining, and the Arctic summer night was almost as light as day, and deep into the night we followed her. Down she went, for the sixth or seventh time, but fatigue was getting the better of her. She was weakening, while,

with all the fatigue, our spirits and strength too were kept up by the excitement. At last, when we had been nearly twenty-four hours on the chase, I got another harpoon in her. This seemed to madden her afresh. Another plunge, which had nearly carried us with her; but this time she did not stay down more than ten or twelve minutes. Up she came once more, the water all around covered with blood, and we knew she was done for. Three or four lances were hurled into her ponderous bulk, and at last our exertions were rewarded by seeing her roll over on her side. She was dead. We bent on another strong line, and soon towed her to a floe. But we found ourselves, with our prize, a good nine miles from the ship. We could not, therefore, save the blubber, but we made a good haul of baleen, with which we loaded our boat to its utmost capacity, and then dragged her, with her heavy cargo, the whole distance over the ice to the ship, which is what I call a fair day's work.

Sailors have a rough life of it, but they often contrive to amuse themselves in circumstances which most landsmen would consider very miserable. ... One day I saw a messmate fixing a lot of strings about six feet long, to the ends of which he affixed a bait of seal blubber; then tying all the strings together at the other end, and also across the middle, he flung the baited ends overboard. Presently a lot of mollimokes espied the food, and one and another seized a morsel, when, suddenly, jerk went Jack's arm, and out flew the blubber from the beaks of the "mollies." Over and over they tried it, until at last, baffled and disgusted, away they flew. But to return to our voyage.

The fog clearing, we pressed our way along through the broken ice till we got near shore, where we found clear water, and went into harbour, some sixty miles north of Cape Walsingham; but after a few days, finding no whales, we steered for the cape, where we found most of the English and Scotch whaling ships. Here were we also unfortunate, and soon left for Cumberland Gulf. ...

It was in the early part of September 1850, when the *McLellan* arrived in Cumberland Gulf, and there had never been but a few ships in those waters. ...

Captain Quayle, hearing from the Eskimo that early in the spring, before the ships were able to get in the bay, there were plenty of whales there, called for volunteers to go ashore and stop through the winter, concluding that the vessel should now go home and return for the shore party when they came up again next year. Twelve men volunteered, of whom I was one. We took our traps out of the *McLellan*, went ashore, and pre-empted a section of land whereon to build a hut or house. The captain gave us what provisions he could spare; but it was not much, for the vessel had only been provisioned for the usual trip, and the owners had not anticipated that twelve men would require food for eight or ten months longer than was customary. There was very little lumber either that we could get from the ship, so we built the house of stones, filled the crevices with earth and moss, and making the roof by laying poles across and covering these with canvas; inside we built berths, or bunks. Before winter was over we got very short of food, and could not have survived if it had not been for the game we shot and the seals we caught. We had to learn the Eskimo ways of eating and cooking, and before spring I was pretty well acclimated; and though the life was so rough and so different to what I had been accustomed to, having lived all my previous life in New York city, yet my health was good; in fact, the whole party kept well.

We had not many opportunities of making pets of any thing out there; the dogs were too fierce, and small animals of any kind were scarce; but one day I saw a young seal; it looked so pretty, with its pure white coat (the young of the Greenland seal [harp seal] is entirely white) and bright hazel eyes, that I took it up in my arms like a baby, and carried it along, talking and whistling to it by the way. The little creature looked at me, turning its head round to look up in my face without any apparent alarm, and seemingly soliciting me to give it something to eat. I thought I should take a great deal of comfort with my little pet, for I had not then got accustomed to seeing the young ones killed, much less eating them myself.

Arrived at our house, I carefully deposited it outside in

a suitable place and went inside to get my supper, hurrying through my meal to get out and look after my treasure. I looked around, but it was not where I had left it. I began to suspect mischief, and sure enough, there it was, a little way off, *dead*, with its back broken by the heavy heel of a whaler's boot; one of the men, with a malignancy impossible for me to understand, had pressed the life out of my only pet simply to gratify a brutal nature. Had I been quite sure who was the perpetrator, my indignation would have found other vent, I suspect, than words.

In the spring we had the satisfaction of knowing that we had not wintered there in vain, as we killed seventeen whales; and, had we been more experienced, we could have captured many more; but this was the first season that any whalemen had passed the winter in that region, and we had every thing to learn.

As summer approached, we began to look anxiously for our ship. All our original stock of provisions had been long consumed, and we had to hunt hard to get enough to eat; and I scarcely believe we should have succeeded in securing enough to sustain so large a party if it had not been for the help of the friendly Eskimo.

While we were busy whaling in the spring, and before we had learned to eat whale meat—for whalemen only strip off the blubber, and abandon the carcass (having also taken the vauable portions of the bone)—the natives would seize upon the latter and strip off all the meat. What they could not eat they put in sealskin "drugs," or bags, and these they stowed away for future use, hiding the bags by covering them up on the various islands in the gulf or inlet. Subsequently, in our hunting excursions, we often came across these "drugs," and if our chase had been unsuccessful, and ourselves very hungry, as was frequently the case, we helped ourselves to these reservoirs of old whalemeat; and as much of it had been lying under the stones for several months, it was not particularly savoury; but we were often very glad, indeed, to get it.

It was not until the month of September—a whole year

having passed—that we were rejoiced by the sight of a vessel. On boarding her, we found that she belonged to Hull, in Yorkshire, England, and was named the *Truelove*; her captain's name was Parker. She had formerly been a privateer in the American war of the Revolution, and was at the time I speak of about *ninety years old*; and the good old bark was still afloat but a few years ago, and Captain Parker was still in her as late as 1860, and is nearly as old as the vessel. She has since been lost. The fact is, no vessel will last so long as a whaler, unless accident destroys her; for once get a ship soaked with whale oil, and it is impossible for her to rot.

On board of her we were surprised to find our old captain, Quayle. He had lost the *McLellan* in Melville Bay; and having put his crew on board of different whaling ships, and sent them home, *via* England, he, with his boat's crew, was taken up by the *Truelove*, Captain Parker kindly consenting to come round and pick us up too; and right glad we were to get a good keel under us again, and some civilized food to eat. But still we could not get home. The *Truelove* was bound for Hull, and on the fourth of October sailed for England. Nothing worthy of special notice occurred on the voyage until we reached the Scottish coast, where we encountered a terrific gale, which the good old *Truelove* weathered; but another whaler which was in our company went ashore and was lost.

T HE *Truelove*, IN WHICH TYSON SAILED to England, was without doubt the most long-lived of all arctic whaleships. Built in Philadelphia in 1764 she made her seventy-second (and last) whaling voyage in 1868, and then served as a merchantman for a decade or more, carrying various cargo around the North Atlantic region, including coal, cryolite ore, petroleum, timber, resin, turpentine, and even ice. In one voyage she reached Philadelphia, the place of her birth 109 years earlier, and was greatly fussed over by the Americans.

Tyson continued his roundabout trip to the United States

Ten ships beset in Melville Bay in July 1852, among them the American whaler McLellan, *which was subsequently crushed and lost. Some of the* McLellan's *crew from her previous voyage (George Tyson among them) were at the time waiting to be picked up in Cumberland Sound, having successfully passed the winter ashore. (The MIT Museum)*

in December on board an American emigrant ship, the *Charles Holmes*, sailing from Liverpool to New York. This experience turned out to be worse by far than the whaling and wintering experiences of the previous year. For forty days the ship beat her way through Atlantic headwinds, with many of her three hundred passengers sick or dying, until the captain gave up and returned downwind, reaching Ireland in a week. Tyson stayed with the vessel during repairs, departed again in March, and arrived home at last in April, having taken more than half a year to travel from Baffin Island to New England.

Having had such a hard experience, and my friends strongly urging the point, I concluded to give up going to sea, and returned to my old business in the manufacture of ironware, but very soon grew tired of it, and again longed for the sea. It has its hardships, but it has its compensations too: at least I was sure that I could never spend my life in the stifling atmosphere of an iron factory; and so, in the spring of 1855, I went again to New London, and shipped as "boatsteerer" in the bark *George Henry*, Captain James Buddington (uncle of Captain S.O. Buddington, sailing-master of the *Polaris*).

On arriving once more on the scene of my old adventures off the entrance to Cumberland Gulf, where we were bound, we encountered an extraordinary heavy pack of ice. ... So, to pass away the time until the ice cleared away, we sailed for Disko Bay, where we were pretty sure to find the "humpback" whale, which we did, making a good catch. In August we sailed again for Cumberland Gulf expecting, of course, by that time, that the pack would be gone; but, to our surprise, it was still there. Never in all my experience have I seen any thing equal to it; but, forbidding as it was, we must "take it" to get into the gulf, though it was so compact and heavy that the July and August suns seemed to have made no impression upon it. But "nothing venture, nothing have." We took the ice off Cape Walsingham; and on penetrating the pack about forty miles, it closed on us, and we were regularly "beset," our drift being to the southward.

In the latter part of August I sighted a vessel, which at first we all supposed to be a whaler, as we knew there were several trying to get in the gulf. This vessel remained in sight several days. At times we imagined she had all sail on, and was working through the ice. No one for a moment thought that she was an abandoned vessel, but there was something about her which aroused my curiosity; I seemed to feel that there would be a story to tell if I could only get at her; and when she had been in sight about two weeks I asked the captain for leave to go over, with two or three companions, to see what she was made of. He objected at first; thought "we should never get there" (she was about ten or twelve miles off); and if we succeeded in reaching her he was sure "we would never get back"; but I was determined, and so at last, in company with the mate, John Quayle, the second mate, Norris Havens, and Mr. Tallinghast, a boatsteerer like myself, we started off for the phantom ship.

It was early morning when we left the *George Henry*, for we knew we had at least ten, and perhaps more, miles to walk. The task we had set ourselves was no light one; the pack was very rough, and every little while we came to patches of open water; and as we had no boat with us, we were obliged to extemporize a substitute by getting on small pieces of ice and making paddles of smaller pieces; and thus we ferried ourselves across these troublesome lakes and rivers. We were all day on our journey, it being nearly night when we reached the stranger. As we approached within sight we looked in vain for any signs of life. Could it be that all on board were sick or dead? What could it mean? Surely, if there were any living soul on board, a party of four men travelling toward her across that hummocky ice would naturally excite their curiosity. But no one appeared. As we got nearer we saw, by indubitable signs, that she was abandoned.

THE DERELICT VESSEL was HMS *Resolute*, which had left England in 1852 as part of Sir Edward Belcher's squadron of five ships sailing in search of Sir John Franklin and his men. Since the voyage of Robert Goodsir on Penny's whaleship *Advice* in 1849 (chapter 8) a number of search expeditions had followed up the efforts of Ross, Bird, Moore, and others. The discovery in 1850 that Franklin and his crews had spent their first winter of 1845–46 on Beechey Island on the north coast of Lancaster Sound, and the subsequent finding by the whaleman William Penny of a piece of wood apparently from HMS *Erebus* or *Terror* in the northern part of Wellington Channel, had turned attention away from the real area of the Franklin disaster. The clues led northward from Lancaster Sound; the minds of the Admiralty and the ships of the search expeditions swung in that direction. In 1852 Belcher took to himself the penetration of Wellington Channel, which looked so promising, while he sent two of his ships, HMS *Resolute*, under Henry Kellett, and HMS *Intrepid*, under Leopold M'Clintock, westward through Barrow Strait. Ironically it was the latter two ships that accomplished the most, duplicating Edward Parry's feat of reaching Melville Island in one season, wintering, and then discovering and rescuing Robert McClure and the men of HMS *Investigator* who, unknown to anyone, had been inextricably beset by pack ice on the north coast of Banks Island for two winters after coming into the Arctic by way of Bering Strait.

But when *Resolute* and *Intrepid*, with the men of HMS *Investigator* on board, headed west in the summer of 1853 to rejoin the rest of Belcher's squadron at Beechey Island, they were overtaken by freeze-up near Bathurst Island and had to pass their second winter in the ice of Barrow Strait. In the spring of 1854, supposing that the ice was not going to release the *Resolute* and *Intrepid*, or the *Assistance* and *Pioneer*, which were then ice-bound in Wellington Channel, Sir Edward Belcher gave one of the most shocking orders ever issued – to abandon all four ships in the Arctic. Their crews, and the men of the *Investigator*, were to make their way over the ice to Beechey Island and travel home on the *North Star* and two other supply ships expected out from England. This incredible decision, unparalleled in arctic history,

H.M.S. Resolute *is abandoned in the spring of 1854;* Intrepid *is in the background. (Metro Toronto Library)*

did not improve Belcher's popularity at the Admiralty.

Before abandoning HMS *Resolute* on 13 May 1854 Captain Kellett and his officers had assembled in the cabin, raised glasses of wine, and drunk a farewell toast to the ship. They could not suspect then that after their departure the vessel would undertake a heroic voyage on her own. With no crew to man sails, sheets, backstays, and braces, she would make her way without mishap through Barrow Strait, Lancaster Sound, Baffin Bay, and Davis Strait for 1,300 miles in the dangerous company of ice floes and bergs to this chance rendezvous with the whaler *George Henry* more than a year later.

By this time Mr. Quayle was so tired that I had to assist him in boarding the ship, myself and the other two following. We found the cabin locked and sealed; but locks and seals did not stand long. A whaler's boot vigorously applied to a door is a very effective key. We were soon in the cabin. This was no whaler, that was plain; neither was she an American vessel, it was soon discovered. English, no doubt

of that. Every thing presented a mouldy appearance. The decanters of wine, with which the late officers had last regaled themselves, were still sitting on the table, *some of the wine still remaining in the glasses*, and in the rack around the mizzenmast were a number of other glasses and decanters. It was a strange scene to come upon in that desolate place. Some of my companions appeared to feel somewhat superstitious, and hesitated to drink the wine, but my long and fatiguing walk made it very acceptable to me, and having helped myself to a glass, and they seeing it did not kill me, an expression of intense relief came over their countenances, and they all, with one accord, went for that wine with a will; and there and then we all drank a bumper to the late officers and crew of the *Resolute*.

It was now too dark to attempt to travel back that night over the broken ice, and we prepared to stay where we were. Possibly the wine we had taken, being at that time unused to it, partly influenced us to this conclusion; but sleep in the vessel we did.

In the morning we found it snowing and blowing very

heavy from the south-east. We could not hope ever to find our way back to the *George Henry* in such a storm, and so, having made a fire, we were prepared to pass the time as comfortably as possible. Among other things, we found some of the uniforms of the officers, in which we arrayed ourselves, buckling on the swords, and putting on their cocked hats, treating ourselves, as *British officers*, to a little more wine. Well, we had what sailors call a "good time," getting up an impromptu sham duel; and before those swords were laid aside one was cut in twain, and the others were hacked and beaten to pieces, taking care, however, not to harm our precious bodies, though we did some hard fighting – *we, or the wine!*

The storm continued for three days, during which we had ample time to investigate the condition and inspect the contents of the good ship *Resolute*. We found food on board, and were enjoying ourselves so well that we should not have cared if it had lasted six. But the weather cleared up, and we saw that the *George Henry* was still at about the same distance from us; so we took all we could carry on our backs, and started to return, arriving at our ship all safe, though some of us got a good ducking by jumping into the water while attempting to spring from one piece of ice to another. Being so heavily laden, we often fell short of the mark, and went plump into the water; but we were in such good spirits that these little mishaps, instead of inciting condolence, were a continual cause of merriment.

On arriving at the *George Henry*, we made our report to Captain Buddington, describing our treasure-trove in glowing terms. After a good rest, we again started for the *Resolute*, and stayed several days on board. At this time the two vessels were nearing each other – the one voluntarily, the other drifting, as she had already done, for a thousand miles. We did not know this at the time, but learned afterward that the *Resolute* had been abandoned, by Sir Edward Belcher's orders, on May 15, 1854, near Dealy Island, and had drifted all the way to Cape Mercy.

At last the two vessels were only about four miles apart.

We were still having a nice time, when one morning, we saw several persons coming over the ice, and, to our discomfiture, they proved to be the captain, with several of the crew. We very soon get orders to return on board the *George Henry*, while, to our chagrin, the captain took possession of the *Resolute*....

THE *Resolute* APPEARED to be reasonably intact so Captain Buddington resolved to sail her to the United States with half of his own crew. Despite deteriorating rigging, rotten sails, and insufficient ballast, they accomplished the feat, navigating with a watch, a quadrant, and a chart sketched out on a piece of paper. During the two-month voyage to New London, Connecticut, they encountered the British whaler *Alibi*, and Buddington consigned to her captain a small parcel intended for Captain Kellet, containing a pair of his epaulettes left behind on the *Resolute*. The package eventually reached him, and it must have been with some fascination and amusement that Kellett, who was by this time on naval duty in the tropics, unwrapped his old badges of rank, in unspoiled condition, years after he had abandoned them with his vessel in the arctic ice.

After the arrival of the *Resolute* in New London the American government conceived the idea of returning her to Great Britain as a gesture of goodwill. Congress allocated the necessary funds to purchase the ship, and the Navy carried out the work of restoring the weather beaten vessel to her original condition. Attentive to the smallest detail they even placed on board such personal items as officers' books, pictures, and music boxes. In November 1856 the *Resolute* was back in her element, driving across the Atlantic under sail to a grateful – if slightly embarrassed – reception in England.

Years later, in 1879, when the *Resolute* reached the end of her useful life and had to be demolished, a desk was made out of some of the wood and sent by Queen Victoria to the American president Rutherford Hayes, who installed it in the oval office of the White House. There it has remained, with only occasional absences, through the last century. When the presidential desk, in its turn, finally reaches retirement age and has to be broken up, its wood will likely be made into something smaller — a chair perhaps — to be presented back to Britain. It has the makings of an endless story.

George Tyson sailed again to the Arctic in 1856, 1857, 1859, 1860, and 1863, wintering on four of the voyages. "I had now become so accustomed to the northern climate," he wrote, "that it seemed more natural to me than a more southern one." In the United States and later near Frobisher Bay he met Charles Francis Hall, then beginning his arctic explorations. As the ensuing narrative shows, he was to meet Hall several times before finally consenting to join him in the ill-fated *Polaris* expedition a decade later.

Sailed again, in the spring of 1864, in the bark *Antelope* of New Bedford, and on this voyage stayed out two winters — one in Hudson Bay, and one in Cumberland Gulf. On this trip I took my vessel farther north in Hudson Bay than any of the whalers had been before. I sailed right ahead into Repulse Bay [at the base of Melville Peninsula], and *took the first whale there that was ever caught in those waters....* There is a peculiarity about this locality which I have never found elsewhere so near the Arctic Circle, and that is the frequency of thunderstorms accompanied by vivid lightning.

While I was in winter quarters in Hudson Bay, Captain Hall visited the bark *Monticello*, which had brought him out, and also other vessels wintering there, including the *Antelope*. I then had long talks with him about getting up another expedition after he had found out all he could about Sir John Franklin's expedition, and he always wound up by saying he wanted me to go with him. He was badly off for boats [whaleboats] at that time, and I let him have one of mine. The *Antelope* was lost in a severe storm in the year 1865, and I returned to St. Johns, Newfoundland, in the steamer *Wolf*, Captain Skinner, and from there got home.

Charles Francis Hall. (Public Archives Canada, C-5913)

Sailed again, in the spring of 1867, in the topsail schooner *Era*, on which voyage the schooner broke out of winter quarters in December, and drifted out to sea. We had two vessels in company caught in the same drift; one was abandoned, the other run ashore. The *Era*, finally drifting in among some bergs, was frozen in for the winter. During this voyage I met Captain Hall again. He was living with the Eskimo in 'training' as the sportsmen would say, for the great work which he even then had in mind. I supplied him with provisions of various kinds, and he, when he had opportunity, sent the natives with fresh meat to the ships.

Sailed again, in the *Era*, in the spring of 1869, returning in the fall of 1870.

In referring to my old logbooks, as well as in recalling the events themselves, I find that the experiences of whaling are not essentially different from those of the polar exploring parties—so far, I mean, as the exposures and dangers are concerned. We were in continual risk of getting "beset," and often were closed in, and unable to move for days or weeks, and sometimes compelled to remain and winter, being unable to break out or bore our way through....

[In]...1869–70 the ship [schooner *Era*] remained frozen in until February, and myself and the crew lived ashore in the house or hut we had built with stones and covered with the sails taken from the ship, watching anxiously all the time for a break-up, which might either relieve the ship or crush her to pieces. I could not tell what would happen; but, fortunately, in February the ice began to break, and I got over to my ship, found she was still seaworthy, repaired damages, got our provisions and other articles aboard again, and, getting a lead out, finished my intended trip, making, after all, a very fair voyage.

IN 1870 CHARLES HALL tried to enlist Tyson as sailing-master and ice-pilot for the new expedition he was planning but the veteran whaleman had other plans and so Hall hired Sydney Buddington (nephew of James Buddington who had captained Tyson's second arctic voyage in 1855) to take charge of the ship. Later Tyson's plans fell through and Hall persuaded him to join the expedition as well, promising to "make" a suitable position for him even though the crew was complete.

The North Pole was the objective of Hall's expedition on board the *Polaris* but they would not reach it. Not until another thirty-eight years had passed would a human being stand at the long-sought top of the world where all directions are south, and even then the achievement would fall in the shadow of the acrimonious dispute over the authenticity of conflicting claims by Peary and Cook, the two pretenders to the throne of discovery.

The *Polaris*, with George Tyson on board, left Washington on 10 June 1871, spent two weeks in New York, and then stopped for a few days at New London, Connecticut, to hire an engineer.

At New York, Hall wrote encouragingly to the Secretary of the Navy, "the officers and crew have taken hold of their work with energy and exemplary conduct," but by the time he reached New London the news was not so good: the carpenter had been consigned to hospital; the steward had been "discharged for incapacity"; and the cook, a fireman, and a seaman had deserted. These were, perhaps, early symptoms of a deep discontent and tension that would in time undermine the entire purpose of the expedition.

Later events would demonstrate that few, if any, of the men on board the *Polaris* shared Hall's zeal for northern exploration, and that his single-minded determination to attain the Pole, along with his high morals and strict discipline, made ordinary men uneasy. There were other sources of friction among the crew. Buddington had been appointed sailing and ice master—effectively the captain of the ship—but he was under the ultimate authority of Hall, the commander of the expedition, who had no marine qualifications. Dr. Emil Bessels from the University of Heidelberg had been put in charge of the scientific program, but he also had to bow to the supreme authority of Hall. This irked the arrogant German. The existence of two whaling captains on the same ships did not promote harmony. Sydney O. Buddington was a secret drinker who openly criticized Hall among the crew, derided the object of the expedition, and let the men do pretty much as they pleased. Tyson was the more competent and enterprising of the two men, but his ambiguous role as "assistant navigator" made it difficult to exercise any real influence on events. He resented Buddington's incompetence and tactlessness, and Buddington doubtless resented Tyson's unspoken criticism and disapproval. As if these were not sufficient ingredients for shipboard tension, there was also a lack of sympathy between seamen and the scientists, between Americans and Germans, and between whites and Eskimos. It was not a happy ship.

The *Polaris* proceeded north to Davis Strait, crossed Baffin Bay, entered Smith Sound, and ran northward for 300 miles between Ellesmere Island and Greenland to a latitude of 82°N before encountering heavy ice. Hall wanted to press on, and George Tyson supported his initiative, but Buddington had no appetite for further adventures, and opposed the plan. Hall gave in, and the expedition took up winter quarters in a bay, which Hall named Thank God Harbour. As it turned out his thanks were premature. In late October, after returning from a two-week sled trip to the north, Hall drank a cup of coffee and quite suddenly became ill. He took to his bed, and put himself under the care of Dr. Bessels, surgeon of the expedition. Two weeks later he was dead. The cause of death, according to Bessels, was apoplexy, possibly combined with a stroke. Almost a century later, however, after Professor Chauncey Loomis and Dr. Frank Paddock obtained samples of fingernails and hair from the icy corpse in Greenland, the analysis at Toronto's Centre for Forensic Sciences showed that Hall's body had absorbed large quantities of arsenic during the last two weeks of life. The person most likely to have administered the poison was Dr. Bessels.

With Hall put to rest there was less pressure to reach the Pole, and the men turned their attention to the immediate problems of wintering. In the following spring (1872) they made only half-hearted attempts to get farther north. The *Polaris* was beset at Thank God Harbour through July, extricated with difficulty in mid-August, and then caught in southward-drifting pack ice for two months, during which she was carried 300 miles south into Smith Sound. On the night of 15 October a powerful gale broke up the ice and drove the floes against the hull with terrible force. Amid the confusion and near-panic of abandoning ship the *Polaris* was suddenly carried away into the darkness with half her crew on board, leaving Tyson and eighteen others marooned on an ice floe.

The situation appeared almost hopeless. They were on a piece of floating ice a few miles in circumference with the arctic winter ahead of them. Like the crews of the whaleships that from time to time had been caught up in the Baffin Bay pack, they faced the prospect of a long southward drift towards the open water of the North Atlantic, at the leisurely pace of the current-propelled floes. But there was a difference. They were commencing their journey at 78°N, more than 300 miles north of the latitude of the *Dundee* and *Dee* when beset in 1826 and 1836 (chapters 2

and 6) and almost 800 miles farther north than the *Viewforth* when trapped in the ice in 1835 (chapter 5). Furthermore, they lacked the security and protection of a ship's hull; they were totally exposed to the elements.

Aside from Tyson, who assumed command of the party, there were nine Eskimos and nine whites in the group. The Eskimos included Hall's great companions and friends from Baffin Island, Ebeirbing ("Joe") and Tookoolitoo ("Hannah") with their child, as well as a Greenland couple with their four children, the youngest of whom was only two months old. The whites, most of whom were German-Americans, consisted of the expedition's meteorologist, cook, and steward, and six seamen. From the *Polaris*'s supplies on the disintegrating ice floes the party had managed to salvage two boats, two kayaks, a tent, a number of guns, a substantial amount of ammunition, some extra clothes, a few scraps of lumber and canvas, and at least six sled dogs. They had also obtained a formidable quantity of food, without which they would almost certainly have perished in short order: approximately 1,500 pounds of ship's biscuit, 630 pounds of beef pemmican, 200 pounds of canned meats and soups, fourteen small hams, twenty pounds of mixed chocolate and sugar, twenty-two pounds of dried apples, and a little frozen seal meat. These provisions amounted to about 2,400 pounds. Estimating that they might be as much as six months in the ice Tyson established a daily ration of twelve pounds a day for the little community, with biscuit and meat in roughly equal proportions. Each adult would receive eleven ounces of food per day and the children half that amount. This was the basic diet; any game secured by hunting would be extra, and would be shared by all.

It must have been clear, at least to Tyson, that the supplies from the ship, godsend though they unquestionably were, would not be enough to keep everyone alive during an arctic winter. Their survival would depend to a great extent on what they could obtain by hunting. But the prospects of encountering animals more than a hundred miles from land in the Middle Ice of Baffin Bay were dismal. And if any animals did happen to be sighted there were still the obvious difficulties and hardships of

A picture of Eskimo Joe from the journal of Hall's second expedition (1864-69).

The separation of the ice-floe party (including George Tyson and Eskimo Joe) from the expedition ship Polaris *during a gale. The*

sketch is by Emil Bessels, who may have poisoned Hall. (National Archives and Records Administration, Washington, D.C.)

stalking, shooting, and harpooning in darkness and intense cold.

The two Eskimo men, Joe and Hans, built several snow huts for the party and then concentrated on hunting. Within a few days they had killed three seals. It was a most encouraging sign; with some fresh meat on hand and a quantity of blubber for makeshift lamps the situation looked a little brighter.

On the other hand, it soon became apparent that the sailors were not to be counted upon as hunters. Although all of them were armed none made any serious efforts to hunt. They preferred to spend their time sleeping and playing cards, or swaggering about with their guns accomplishing nothing. Like

the five children in the party they would have to be reckoned as consumers rather than producers of food during the winter months ahead. Not only did the sailors become lazy and dependent but they grew insubordinate as well; they ignored Tyson's authority, frustrated his sensible and fair procedures for the sharing of game, pilfered from the food stores, stole guns and clothes, and secretly planned to take a boat (the *only* boat left, for they had wantonly destroyed one for firewood, disdaining to use blubber lamps for their cooking) and set sail for the Greenland coast. They contributed nothing to the survival or welfare of the group, but expected all the while to receive their part and more, not only of the rations saved from the *Polaris* but also of any seals killed by the two Eskimo men, who hunted incessantly in appalling conditions to secure sustenance for the ungrateful wretches as well as for their own families. As the months passed Tyson became increasingly worried that hunger might induce the men to take desperate measures to obtain food, yet he dared not assert his authority too strongly in any matter. They were armed, numerous, irrational, and inexperienced in the Arctic; a showdown could easily end in disaster for the entire party.

The small ice platform continued to float slowly through Baffin Bay in company with its numberless consorts of floes and bergs. Upon its surface, in squalid snow huts, nineteen men, women, and children in tattered and insufficient clothes hung on tenaciously to life in the dreadful cold.

It was the heroic efforts of the two Eskimo men—Joe in particular—that weighted the scales on the side of survival. They secured eleven seals in the severe and barren months of October through January. Then, fortunately, wildlife gradually became more abundant. In early February narwhals were seen in the leads and pools within the pack. The hunters shot ten but none could be retrieved before sinking (the sailors had earlier dismantled the only whaling harpoon). Nine seals were killed during the month, however, and more than fifty dovekies—the first birds taken during the winter. In March the game catch was even greater: nineteen seals; eight hooded seals; one bear; and almost eighty dovekies. In April another bear and three more hooded seals were added.

By this time the floe was off the coast of Labrador, and the influence of the open Atlantic began to make itself felt upon the pack, eroding the size of the ice fragments by collision and abrasion, and making their little base unsafe. On 1 April they abandoned their floe, crowded all nineteen people and as much food and equipment as possible into a boat designed for half a dozen men, and made their way precariously away from the open sea, deeper into the protective ice, camping on floes for as long as they were secure, then moving on farther to the west. On the night of the twentieth, gale-driven seas broke repeatedly over the floe on which they had settled, carrying away their tent and most of their belongings; only by holding on to the precious boat until daylight were they able to save it. A few days later they moved on again, rowing for eight hours before taking up residence on yet another floe, and wondering all the time when and how the desperate ordeal would end. Suddenly spirits soared; a ship was seen! But she went off into the distance without having sighted them. On the next day hope revived again when another vessel was spotted, but the same unbelievable thing happened. Finally, on 30 April, the Newfoundland sealing steamer *Tigress* sighted the figures on the ice and took them on board.

The survival of the *Polaris* castaways was a magnificent feat of human endurance. Nineteen people had lived on arctic pack ice for six and a half months, drifting with the currents more than 1,700 miles from 78°N to 53°35′N during the coldest part of the arctic winter. In a region thought to be devoid of wildlife they had obtained food. Not one person had perished. Even the infant Charles Polaris, only two months old at the start of the voyage, had lived through the experience, on an unorthodox baby diet of blubber, seal offal, pemmican, pulverized hardtack, and brackish water.

Departure of the whaling fleet from Dundee.
(Dundee Museums)

Thirteen

You May Never See Your Thornton Again

... we steamed up the inlet, ... right up into the Gulf of Boothia, past the Bellot Strait, ... a great deal farther than ever whalers reached to before, ... THOMAS THORNTON MACKLIN, 1874

STEAM ENGINES AND SCREW PROPULSION were cautiously introduced to some ships of the British whaling fleet in 1857 to supplement their normal sail power. The combination of steam and sail produced either steam whalers, such as the *Narwhal* of the selection below, or steam tenders, such as the *Isabel*, which assisted the sail whaler *Emma* in 1859 (chapter 10). There were great expectations for these technological innovations, because whales were getting scarcer and catches were diminishing on the Davis Strait grounds; something was needed to give the whalemen a greater advantage over the whales than they had previously enjoyed. The industry had declined alarmingly since the boom days of the 1820s; British ports had sent 742 ships to the Davis Strait fishery between 1820 and 1830, but in the decade 1850–60 they dispatched only 166.

Arctic whaling was suffering from a grievous self-inflicted wound. By pursuing the short-term goals of maximum catches every season the industry had undermined its own resource base; it had destroyed the stock of whales upon which it depended. There had never been any control over the number of companies participating in the whaling, the number of ships operating on the grounds, the number of whales killed per ship, or the duration of the whaling season. The expansion of whaling to the West Side in 1820 had been amply rewarded by incredibly high yields (on average of almost eleven whales per vessel during the next decade) but these rewards were achieved at a tremendous price – a drastic reduction of whale population. The unfortunate legacy of this utterly foolish policy of "raping the resource" was a pitifully small stock of whales, which could tolerate very little further exploitation. By the decade following 1900 the average catch per ship had fallen to one and a half whales, only a seventh of what it had been eighty years before.

After the peak catches of the 1820s the whaling industry spent the rest of the century struggling to adapt to the inevitable biological fact of fewer and fewer whales in every decade, while stubbornly retaining its suicidal policy of unlimited killing. The principal response to whale scarcity was economic withdrawal; companies, ships, and men simply went out of whaling. As many as ninety ships had sailed to Davis Strait in the best years, but the maximum during the decade 1900–10 was only eight vessels in a season. Despite the reduction of whaling effort, average catches continued to drop. Efforts were made to increase the effectiveness of whaling ships and crews: new areas were exploited; longer voyages were attempted; the practice of wintering was adopted; shore stations were established; Eskimo whalemen were employed. Catches declined nonetheless.

The procedure for catching whales in the nineteenth century was not fundamentally different from that employed by Eskimo hunters of the Thule culture a thousand years before. The Eskimos had pursued whales in open skin boats (*umiaks*) about the same size as wooden whaleboats; they had thrown toggle-headed harpoons into the animals to penetrate the skin and hold fast, and had made the kills with lances. They had used the same principle of weakening and tiring the whale by creating a drag on the harpoon line, with the difference that the Thule whale-men attached inflated sealskin floats to the line while the European and American whalemen used their whaleboats as drogues, remaining connected to the whale and playing it, as a sport fisherman plays a fish, by hauling in or slackening off the line.

The occasional introduction of new weapons and the refine-ment of old ones improved the effectiveness of arctic whaling but not to a revolutionary extent. Harpoon guns had been widely adopted in the British fleet; toggling harpoon heads, shoulder guns, and darting guns were to become useful during the second half of the nineteenth century, but the traditional hand-thrown harpoons and lances were never entirely dis-carded. Much was expected of iron hulls when first used in 1857 but the loss of three iron-hulled whalers within a few seasons, including the large Peterhead-built *Empress of India*, which carried eleven whaleboats and more than a hundred men, cooled the enthusiasm of their supporters.

Steam propulsion, on the other hand, gave the whalemen undoubted advantages. Auxiliary engines added to existing or new sailing vessels were very useful against adverse winds and currents, or in calms. Propellers, which could drive a ship forward or backward, added manoeuvrability as well as power. Steam whalers had a greater capacity to bore through ice and, by using reverse gear, to extricate themselves from ice when stuck

fast. They could get to the whaling grounds faster, move more rapidly through the pack to areas of whale concentration, explore remote inlets and bays with greater security, and remain in the Arctic longer than usual in the autumn. Steam auxiliary whalers quickly proved to be better than simple sailing vessels, and soon came to dominate the British arctic fleet. They were not impregnable, however, and if they did have the misfortune to get caught in the ice they faced the usual risks.

The idea of applying steam power to arctic ships was not a new one. Thirty years earlier John Ross had been keenly enthusiastic about using steam engines on exploration vessels, and in 1829 he had persuaded the Admiralty to install one in the ship *Victory*, a small steam paddle-wheeler, to assist her in the search for the Northwest Passage. Boiler leaks had developed soon after departure, and the state of marine engineering in those days is perhaps indicated by the instructions then provided by the firm of Braithwaite and Erickson, the engine's makers, which advised that leaks could be plugged with a mixture of dung and potatoes. They did try this extraordinary concoction but the remedy failed. And when the "patent contrivance," as Ross called the engine (he may have used more explicit terms on occasion), did work it proved to be no more powerful than two boats under oars towing the vessel. Furthermore, the engine, boilers, and coal supply made up no less than two-thirds of the *Victory*'s tonnage. When the expedition went into winter quarters in Prince Regent Inlet Ross ordered the men to dismantle the engine and boilers and discard them on the beach.

In 1845 Sir John Franklin gave steam power a second chance, but with screw propellers instead of paddle wheels. The *Erebus* and *Terror* were provided with old railway locomotive engines of twenty horse-power, designed to drive the vessels along in good conditions at breathtaking speeds of almost four knots. What use was made of this auxiliary power before the two ships were crushed by ice near King William Island cannot be known. Steam vessels were subsequently used on the expeditions of Austin (1850–51), Belcher (1852–54, chapter 12), and M'Clintock (1857–59).

In 1874 only fourteen British and two American ships went to the Davis Strait grounds. The American vessels, both sail-propelled, probably operated in Cumberland Sound, but at least twelve of the British whalers, all of them steamers, took the traditional route up the Greenland coast. One of these ships, the Dundee steam whaler *Tay*, was crushed by ice in Melville Bay but the others made it safely to the West Side on 4 June (a month earlier than the usual date of arrival for sailing ships) and spent the next month cruising around the mouths of Pond Inlet and Lancaster Sound. They secured a total of fifty whales and then steamed westward through Lancaster Sound and headed south into Prince Regent Inlet.

This region had been visited by several exploring expeditions and they had reported whales there. As long as the whaling was good off Pond Inlet, however, there was not much incentive to take a sailing ship 300 miles farther west against opposing current flow and ice drift. A few made the voyage, nonetheless, including the Hull whaler *Isabella* as early as 1834, but the real popularity of Prince Regent Inlet awaited the age of steam. Engines added enough security and speed to make the venture more attractive.

One of nine steam whalers entering Prince Regent Inlet in early August 1874 was the *Narwhal* of Dundee. Her surgeon, Thomas Macklin, like other medical men such as William Cass (chapter 1), the anonymous doctor of the *Hercules* (chapter 3), John Wanless (chapter 4), Alexander M'Donald (chapter 7), and Robert Goodsir (chapter 8), was an alert observer and a methodical recorder of events and scenes during the voyage. Obviously well-read, he considered the large body of water into which the *Narwhal* was sailing "peculiarly interesting, on account of...several of the harbours in it being the wintering quarters of many of the exploring and discovery expeditions." The voyage south past Port Leopold, Batty Bay, Port Bowen, Creswell Bay, and Fury Beach evoked memories of Parry, John Ross, James Ross, Kennedy, M'Clintock, and other "greats of our modern arctic navigators." Ashore Macklin saw the graves and the debris of their expeditions; on board ship he joined others in

eating plum duff made with flour abandoned on the beach by Parry's expedition a half-century before. He found it "quite good."

With a confidence engendered by steam propulsion the captains boldly took their vessels beyond Bellot Strait but, having no success in whaling, they returned northward, only to find their escape route barred by ice. In attempting to force a passage all the ships were beset. Surgeon Macklin describes the crisis and goes on to relate subsequent events as the *Narwhal* proceeded southward in Baffin Bay to carry out the "fall fishing" around Cape Kater.

Whale-kill symbols in the journal of surgeon Macklin on the Narwhal *in 1874. (Dundee Museums)*

FRIDAY 6 AUGUST

...Alas! I may never see you again, for here we are, hard and fast, among the ice above Fury Point. The ice has squeezed us twice and our stern post is started; all our boats and provisions are on the ice, ready to leave the ship, if she goes. The *Arctic* is gone, though not sunk yet. She is in such a state that the men dare not stay in her. They are all on the ice.

But let me tell you briefly how things are. On Saturday, we steamed up the [Prince Regent] Inlet, heaving to at Fury Beach for a few hours, then proceeding right up into the Gulf of Boothia, past the Bellot Strait, which divides North Somerset [Somerset Island] from the most northern part of America, being a great deal farther up than ever whalers reached to before, and in fact farther than even many explorers have reached. Well on Tuesday last we turned to come back, also all the rest of the fleet, but by the time we got to Fury Point we could get no farther so our respective positions now are: *Narwhal*, nearest the open water (we can see no open water from the masthead, but we are nearest to where we know it to be) and in the "nips" [squeezed by the ice]; *Camperdown* next, not in the "nips," but boats on the ice in case of anything happening; *Arctic*, abandoned; *Intrepid* in the "nips" and boats and provisions on the ice; and *Victor*, not in the "nips" but boats on the ice; away above these three, which are pretty close together, are the *Erik* and *Polynia* together, and farther off is the *Mazinthian*. How the last three are faring we cannot tell, being too far off to be able to make out with the glass. Such a mess, or so many ships to be in such a plight, was never known before in the annals of arctic whaling adventure. Here we are all in a predicament, not one to pick us up. Even if we do get to clear water with our boats, the nearest ships being five or six hundred miles off...we never could reach that distance in open boats.

(The mate has just come to tell us that the *Arctic* is afire; the captain and I went on deck, and there she was blazing away).

We could not carry provisions enough with us for the time it would take us to go that distance, and I question very much if boats would carry us on account of the sea which would most likely be on; and then if we have to winter we will require to go on short allowance, one third the usual quantity;...how many could live on such a regimen as that, during the rigours of an arctic winter? I cannot write more. God be with us. He will preserve us if it be His will. You may never see your Thornton again. His will be done. 11 P.M. Hurrah the ice is easing off. It has been blowing a fearful gale all day, but it is taking off a little now. Alas! Instead of easing it is nipping harder.

THURSDAY 13 AUGUST

The ice has cleared away, and we are now steaming down along the shore, and are now past Fury Beach and still clear water ahead. We may get out now, if there is not a barrier farther down. On Saturday last at 9 A.M. the *Arctic* went down in Lat. 72°30′ and Long. 93°. At night I walked down to the *Victor* to see how they were faring. They had received no damage. I then went over to the *Intrepid* to see the carpenter of the *Arctic*, who had got himself burned, the decks having given way and precipitating him into the flames, from whence he was extracted pretty severely burned. However, he is doing well. On Monday Henderson came over and had tea with us and that same night we got out into clear water, that is, a lead or narrow strip of clear water, and there anchored to the floe. Next morning the doctor of the *Intrepid* came over to the ship and asked me to go and help him dress the burned man. While there the ice began to break up and before I left the ship she had got underway and I was detained a prisoner, and it was rather opportune that I was, for after dinner, while one of the hands was aloft clearing the main topsail buntlines, he got his left hand crushed some way or the other. The ensign was hoisted and dipped to the *Victor* who was close ahead of us, the *Camperdown* and *Narwhal* being far ahead. Henderson came aboard, and getting the man under chloroform, we took off the four fingers, leaving the thumb....

It is a mercy that we are not detained, for if we had required to winter very, very few indeed, would ever have seen home again, as we have only provisions to last till Christmas on full allowance,...I had almost forgotten to mention that we have twenty-one extra hands over and above our own men to feed, namely six of the *Tay*'s crew—already mentioned—and fifteen of the *Arctic*'s crew.

SATURDAY 15 AUGUST

The *Victor* is away home. Yesterday she hoisted her ensign to signify she was going home. She took a lot of the shipwrecked men, thirty-six in all, including the captain of the *Tay*...also Captain Adams of the *Arctic*...and a lot of hands, also the man who got his fingers taken off on board the *Intrepid*. So you see Henderson has a lot of company on board now. There are eighty-two hands, all told, on board the little *Victor*. You would hardly believe she could carry them, if you saw her. At least, I know none who see this would care to cross the Atlantic in her. Yet she is a brave, stout little ship and will take them all safe home, I trust. I sent a letter and a small parcel with Henderson for the folks at home....

MONDAY 17 AUGUST

We are now bound away down south to Cape [Henry] Kater, where we will lie in harbour till the fall fishing commences, which is generally about the middle of September....

FRIDAY 21 AUGUST

We are now a little way beyond Cape Hay and it is dead calm. In fact it had [been] calm since Monday and generally thick in the afternoon till about twelve midnight, when it generally clears up. It has been exceptionally fine weather for Lancaster Sound.

Last night two boats went off to the loomery at Cape Hay, where the cliffs rise perpendicularly out of the sea to about 300 feet high, in some places overhanging; yet the water is so deep at the base of the cliff that the ship would lie alongside without taking the ground. But what birds, chiefly looms, but also some kittiwakes and burgomasters &c. They were innumerable, hundreds, thousands sitting on clefts and ridges on the face of the cliffs. However, we (I was with the second mate) did not waste much time looking at them, but immediately commenced the slaughter. We had just to pick two of the ledges on which we thought the birds were thickest—which was a difficult matter; I fired at the one and the second mate at the other, and while we were reloading the hands in the boat picked up those we had knocked down. By that time the birds that had flown off had returned, only taking a short flight and coming back again. Indeed lots did not take the trouble to fly off, but continued sitting. It may give you some idea of their number when I say that they literally darkened the sky; we could hardly see through them when they flew off. In fact, one time when I had shot three that tumbled right back and lay on the ledge one of the hands in the boat was so provoked that he volunteered to climb up for them, rather a hazardous undertaking when you bear in mind that there was a good heavy swell rolling in, at one time raising the boat high up the face of the cliff and then sinking away down, so that there was great difficulty in landing. However, he managed it and got safe down again. The other boat got about one hundred birds so we have a good stock of fowl now.

While we were away shooting there were some Yacks aboard.

MONDAY 24 AUGUST

For the last two days we have been lying becalmed off Cape Byam Martin.

As idle as a painted ship
Upon a painted ocean.

The weather was beautiful and calm, hardly a breath of wind, the sun shining bright and strong, and the weather mild as a mid-summer day at home. One would hardly realise he was in the arctic regions if it had not been for the snow-capped peaks and deep, dark ravines unclad by heath or moss or the slightest sign of life, and the sea studded here and there with bergs, many of which lay under the shadow of the high and frowning precipices of the Byam Martin range of mountains. I watched as the sinking sun pursued its way to lighten up the dreary shores of the Polynian Sea perhaps for the last time, till another season bring it back to gladden the heart of the Eskimo and to soften the austerities of these cold regions by the rays of its glorious light. I would I were a Sanders, to make a sketch with my pencil, or even a Scott, to paint with my pen, so that I might share with you some of the pleasure which I enjoyed while gazing on that magnificent scene, but it may not be.

THURSDAY 27 AUGUST

For the last two or three days we have been making what way we could against a southerly breeze. Yesterday it fell calm. We got among some ice in the afternoon, where the ship was threatened by a bear. He must have been hungry, poor fellow, for instead of trying to run away he stood his ground, intending, no doubt, to make a meal of the strange black bird with the huge white wings; but flame shot forth from her head and he was brought to his knees in supplication for mercy. But his life's blood was necessary to expiate the crime of his arrogance and intrepidity in daring to approach so mighty a bird; again flame shot forth and the messenger of death demanded of him his life. Such is the punishment which awaits prideful ignorance.

SATURDAY 29 AUGUST

A beautiful, calm and warm, sunny day. Off Clyde River; lots of ice. This morning before breakfast I went away with the second mate bear shooting. We shot three, that is, Mr Anton did, for though I fired twice I missed both times.

We were away from the ships forty minutes only – smart sportsmen.

MONDAY 31 AUGUST

Kater Harbour: we arrived here late on Saturday night. It is a fine, snug place, and if you go well up into it, you are quite landlocked. It is in an island which is separated from the mainland by a narrow fiord. You can see the fiord from our main top, over a low part of the island. There is no game to be seen, but there should be deer, wolves, foxes &c. There are no Yacks here as yet; they are expected soon. The foundations of their winter huts are seen, and certainly they are small enough. There are also a few graves, and also one of a harpooner. Over his grave is a board with the following inscription:

In memory of D Vallance, aged 41, who died on board of the s.s. Intrepid, *on the 26th September, 1871.*

Graves here are vastly different from graves at home. Instead of making a hole six or eight feet deep, a hole is scraped six or twelve inches deep only. Why? Because it is only on the surface that the ground is soft. Beneath it is hard-frozen. The body laid therein then is covered over with stones so as to keep off wild animals from getting to them. The Yacks do this so carelessly that you can see into them. I came across a child's grave – I had one of his limbs in my hand. They [bury?] all their belongings with them, for use in the happy hunting grounds.

WEDNESDAY 2 SEPTEMBER

Yesterday our boats got fast to a fish, while rocknosing, that is, boats which are sent away to the rock [illegible word] outside the harbour. Every ship has away four or six boats while in harbour, with the merest chance of getting a fish. We got up steam as soon as we got word, and went out and towed her into harbour, when we proceeded to flinch her, which operation I will now try and explain.

*Above: Spanning on—splicing a foreganger
onto the sliding ring on the double shank of a
gun harpoon. Photo by Livingstone-Learmonth.
(Public Archives Canada, C-88307)*

*Right: Captain William Adams demonstrating
the position of a lookout in the crow's nest
before it was hoisted to the masthead,
1889. Photo by Livingstone-Learmonth.
(Public Archives Canada, C-88309)*

The fish is always brought alongside with its tail forward abreast of the fore-chains. It is then secured by means of a tackle from the fore rigging, which is hooked to a strop round the small part of the tail by a stout rope, called the "rump rope." A similar purchase is hooked from the main rigging to a strop rove through a hole cut in the extremity of the under jaw, which is called the "nose tackle." The right fin of the fish, which is next the ship's side, is dragged taut up and secured by a chain or rope to the upper deck, the bulwarks on the port side being unshipped.

Between the foremast and mainmast is a stout wire rope, called the "blubber guy," having four large single blocks strapped to it, through which are rove the fore and main spek [blubber] tackles. The former is worked by the capstan and the latter by the steam winch near the mainmast. These tackles are used for hoisting on board the large layers of blubber, some between one and two tons in weight, as they are cut off.

Top: Harpooners pose in a whaleboat suspended from the davits, 1875. (National Library of Scotland)

Bottom: A whale alongside the Scottish steam whaler Eclipse, *ready for flensing. (Dundee Museums)*

From the mainmast head is a heavy purchase called the "kent" or "cant" tackle, which is used to turn the fish over as it is flinched. It consists of a treble and a double block, having a seven-inch fall. Two boats, called "mollie boats" attend upon those cutting up the fish, and are kept alongside it by a couple of hands in each boat, called "mollie boys." These boats hold the spades and knives &c. The harpooners, under the guidance of the speksioneer, are on the whale, and with their blubber spades and knives separate the blubber from the carcass in long strips, which are hoisted in by the fore and main spek tackles. Previous to this, however, a strip of blubber from two to three feet in width is cut from the neck just abaft the inside fin, and this is called the "cant." A large hole is then cut in this band of blubber, through which is passed the strop of the cant purchase, and secured there by a wooden toggle or fid being passed through. By means of this purchase, brought to the windlass, the fish is turned over as required. Each harpooner has iron spikes called "spurs" strapped on to his boots to prevent him from slipping off the fish. The belly is the first part of the whale operated on. After the blubber from this part has been completely taken off and the right fin removed, the fish is canted on to its side by means of the large tackle, and the blubber from the opposite side similarly stripped. The whalebone is then detached, special bone gear being used for this purpose, and the lips hoisted in, and so on till all that is valuable has been cut off and taken aboard. The tail is then separated from the carcass, or "kreng" as it is called, which latter being released generally sinks, and three cheers are given. The duties of the boatsteerer during this operation are to cut up the strips of blubber as they are received on deck, into pieces about two feet square, with long-bladed, hafted knives. The pieces are seized by the line-managers, armed with "pickies," and transported below through a small hole in the main hatchway. Below they are received by the "skeeman" and another man denominated a "king," by whom they are stowed temporarily between decks, until sub-divided into pieces of three or four blades, when what is called the gum, which connects them together, is removed. There are from three to four hundred blades on each side of the head.

The tail is cut up into blocks for chopping up the blubber on, so as to preserve the edge of the cleavers.

"MAKING OFF"

The blubber is hoisted on deck again. It is then seized by two men on each side of the deck, who with their pickies drag it to two men—generally harpooners—stationed on each side, whose duty it is to cut it up into pieces about twelve or sixteen pounds weight, and who remove from it all kreng and other extraneous matter. These men are called "krengers." The blubber is then thrown forward to the remaining harpooners, who are stationed on each side of the deck near a "clash," which is an iron stanchion firmly fixed into a socket in the deck, standing about three feet high, and having five iron spikes on the top.

Each "skinner," as they are called, has an assistant, who is called a "clasher," who picks up the pieces of blubber having skin on it with a pair of clash hooks, and places it on the top of the clash. The skin is then separated from the blubber by the skinner, armed with a long knife. The blubber is then deposited in a heap, called a "bank," in front of the "spek trough," which latter is an oblong trough about eighteen feet long and two feet in width and depth, which is placed over the hatchway through which the blubber is to be passed down. A hole about a foot square is cut in the centre of this trough, to which is fitted a long canvas shoot [i.e., chute] or hose, called a "lull," the end of which is pointed in to the tank receiving the blubber. The lid of the trough is turned back, and is supported underneath by chocks, so as to form a table about three feet high, on which are placed the blocks cut from the whale's tail. Behind the blocks are stationed the boatsteerers, armed with choppers, [whose] province it is to chop up the pieces into small portions after they have

Splitting the slabs of baleen apart to be gummed,
cleaned, dried, and stowed. (Dundee Museums)

such time as an opportunity may offer for the final operation of "making off."

The whalebone, on being received on deck, is split up into portions, each containing from nine to sixteen blades, by means of large iron wedges, and these are again passed through the hands of the skinners. They are then thrown into the spek trough, passed down through the lull and so into the tanks.

The upper deck, indeed, presents a most animated and busy scene during the time the work is at its height. Of course, this tends to make the ship in a most filthy and a greasy state, although there is nothing absolutely repugnant or disgusting in witnessing the process.

THE ABOVE DESCRIPTION of flensing and making off, which comprises the journal entry for 2 September, is virtually identical to one in Albert Hastings Markham's book *A Whaling Cruise to Baffin's Bay and the Gulf of Boothia*. Markham's book was based on a voyage in 1873 on the steam whaler *Arctic*, whose destruction in Prince Regent Inlet during the next season was witnessed by Macklin. The book was published in 1873, presumably before the departure of the *Narwhal*, and surgeon Macklin, like W.E. Cass of the *Brunswick* (chapter 1), was not above borrowing an effective passage to incorporate without acknowledgement in his own journal.

MONDAY 7 SEPTEMBER

Still in harbour. Snow fell on Thursday and Friday pretty heavily, and there is about six inches in depth on the ground, giving a very wintery appearance to everything. A few Yacks have arrived, who say plenty more are coming, as soon as fine weather sets in, with their umiaks or luggage boats. The stragglers are staying on board till the women arrive with the tupoks, or tents.

All the ships have gone out but two, the *Esquimaux* and *Polynia*.

FRIDAY 11 SEPTEMBER

Still in harbour. All the ships have come back, having got nothing during their cruise except bears. *Polynia* went out this morning. A lot more Yacks have come here during the last day or two. There is quite a small village here now, consisting of twelve or fourteen tupuyks [tents], with a population of about fifty individuals, any amount of dogs, and *lots of dirt*.

MONDAY 14 SEPTEMBER

Still in harbour. I have been spending my time in a variety of ways, in visiting the tupuyks and making love to the young cunahs – not the married ones, which would be rather dangerous, for the husband would not waste time in talking to his wife, but would quietly stick a knife in your ribs or do something else equally disagreeable. Some of the young girls are not bad looking, though copper-coloured and not wearing petticoats, their dress resembling that of the men, except that they have a prolongation of the upper garment in the shape of a tail about a foot in breadth, which they put under them when they squat, unlike our fair damsels at home, who lift their appendages (panniers &c.) when they sit down. In one thing they do, however, resemble the girls at home, that is, in giving preference to, or at least receiving the attentions of the upper...rather more graciously than the lower class, in short, preferring officers to men, and sometimes, I think, smiling more sweetly on medical officers than on others. At least I have got a fair share of their endearments, and they are not withheld by excessive shyness on their part, a few fleas and plenty of hairs from their deer skins.

The nursing of infants is somewhat peculiar, the women carrying their cradles on their back, in the shape of a hood, into which the baby is thrown – or put, I should say – when it is not being suckled, and there is rocked and swung about as the woman moves. They are very good-natured and do not cry much, but then they are very fat.

By the bye, that process of suckling the child is very often gone through under very trying circumstances for both mother and child, at least so most of our tender-hearted matrons at home would think, who cannot let the children out with the nurse without half smothering them with clothing. Only last Saturday while climbing the hill on the south side of the harbour, when about halfway up, I came across a cunah giving suck to her child. There was a good breeze blowing at the time too, aye snow on the ground, with the thermometer a few degrees below the freezing point, when my cheeks were blue and my ears nipping, yet there she was quite contented. Evidently it was nothing strange to her; and indeed neither it was, for Captain Hall says he has seen the same thing done with the thermometer standing at forty degrees below zero. All the children are very fat, very interesting, full of fun and frolic, but generally very dirty. The captain tells me one of them, a father, died of grief at the loss of his child.

I also spend a good deal of time in paddling about in a kayak or canoe, but one requires to be very careful when in them, to keep from upsetting. They are so very light, being merely a frame covered with skin. One of our hands was upset three times. If you manage to get out you are pretty safe, for it won't sink though full of water, and will bear you up till succour comes. I can manage them very well now, though I was very nearly over the first time I got into them. If the weather keeps fine I think we will go out soon, and see if we can get a fish or two, for I expect we will bear up for home in the course of a few weeks, three perhaps.

I T WAS MORE THAN THREE WEEKS before the whaleships departed for home. An unusually favourable stretch of weather encouraged the captains to persevere longer than usual in the fall whaling. The ships sometimes steamed out of the harbour with the boats on board and cruised for the day. At other times, on the "warmer" days, the boats were sent out from the harbour to "rocknose" along the coast for several hours; the crew cooked their breakfast and dinner on small stoves and ate in the boats. It was cold, uncomfortable, hard work, and the men were eager to leave for Scotland. A severe gale on 7 October served as a harsh reminder to the masters that winter was not far off. Several ships did not manage to get back into harbour in time and had to spend the night outside in violent winds, heavy seas, and blinding snow. They returned the next morning, frosted like ghost-ships, six of their boats smashed or carried away by heavy seas and a number of hull and deck fixtures damaged. It was quite suddenly a different Arctic; the gale had driven old ice floes down from the north, and young ice was forming swiftly around the shore. Unquestionably it was time to leave. As the vessels departed on the tenth the whales seemed more numerous than ever, enough to set the whalemen's "teeth watering," but no one seriously wanted to linger in those wintry seas.

After a fast homeward crossing of only eighteen days the *Narwhal* made a landfall, and surgeon Macklin rejoiced at the first sight of Scotland. For the young medical student half a year away from home, who had written so candidly in his journal that he had been ashore on Baffin Island "making love to the young cunahs," and who would shortly record a brief but enjoyable flirtation with a Shetland girl, it was perhaps natural to perceive characteristics of womanhood even in the rugged coasts of Scotland. He wrote:

Bonnie Scotland! I once more behold thee, but
like a shy maiden, conscious of her charms,
especially to one who has not had the privilege of
looking upon her for some time, thou hast veiled
thyself in a cloud of mist.

Walter Livingstone-Learmonth on the Dundee steam whaler Maud, *1889. This and all other photos in this chapter were taken by Livingstone-Learmonth. (Public Archives Canada, C-88328)*

Fourteen

The Age of Chivalry

...All this expense and voyage that ladies may wear whalebone instead of steel in their — well, unmentionable articles of clothing. What will not men do for the fair sex! ... They endure arctic frosts and snows in order that they may wear tight dresses with comfort, and yet women cry out for their "rights," and vow that the age of chivalry has passed.
WALTER LIVINGSTONE-LEARMONTH, 1889

IN THE DELIGHTFUL BOOK, *Of Whales and Men*, R.B. Robertson remarked upon the fact that baleen was virtually the only part of the whale discarded in antarctic whaling of the 1950s, whereas a century before in the arctic fishery, it had been reckoned of great value. Indeed, the high value of baleen — popularly called "whalebone" or simply "bone" — was one of the principal reasons why traditional arctic whaling lasted out the nineteenth century, long after the scarcity of the animals should have brought the industry to a halt.

Baleen is the tooth-substitute used by all of the whalebone whales (*mysticeti*) to obtain food. It enables them to extract small shrimp-like krill from sea water by filtering, an entirely different method of feeding than the tearing and chewing of fish and squid that is accomplished by the sperm whale and other toothed whales (*odontoceti*). Within the cavernous mouth of a bowhead whale are approximately 700 slabs of baleen, hanging down from each side of the upper jaw, each overlapping its neighbour. Roughly a foot wide at their base in the gum, the slabs taper towards a point. A fringe of coarse hairs on the trailing edge of each slab catches the krill after water is taken

into the open mouth and then pressed out through the baleen sieve. In contrast to humpbacks, grey whales, blue whales, fin whales, and most other baleen whales, whose baleen plates are usually less than three feet long, bowhead whales may have slabs exceeding twelve feet in length, a characteristic that made the species especially valuable when whalebone was in great demand.

Resilience, lightness, and ease of splitting along its length made baleen an important commodity in the pre-plastic era. Although we may disagree with Livingstone-Learmonth's fanciful notion that whalemen put up with the hardships of arctic whaling so that women could dress fashionably, the fact remains that the demand for whalebone in the late nineteenth century was the most important economic incentive for pursuing the remnants of the Greenland whale population in Davis Strait. Whale oil, which had long been used in street lighting and machine lubrication, was losing ground to petroleum and natural gas; electricity was on the horizon. At the same time the use of whalebone in corsets was expanding. The value of bone steadily outstripped that of oil. According to Professor Gordon Jackson, author of *The British Whaling Trade*, the price of bone rose from £500 a ton in the 1870s to almost £3000 a ton by 1902, and even in Dundee, where the presence of the jute industry created a persistent demand for whale oil required in the softening of the fibres, the value of oil landed in 1905 was less than one eighth that of the whalebone.

On the American market the trend was similar. By 1890 roughly eighty per cent of the value of a bowhead whale was supplied by the baleen in its mouth. Worth about five dollars a pound in 1905, the ton or so of bone provided by one whale could fetch roughly $10,000. So low, by comparison, was the value of oil that some American whaling captains were satisfied to extract the bone and let the rest of a carcass drift away and rot – fifty tons of animal destroyed for one ton of marketable baleen. Now, it seems, the opposite occurs. Everything is used except the baleen, which is hurled over the factory ship's side back into the sea.

Walter Livingstone-Learmonth, from whose unpublished journal the following selection is taken, was not a whaleman by profession. His detailed account of a voyage on the Dundee steam whaler *Maud*, including frequent references to the hunting of animals, reveals a man who had time on his hands and a certain freedom of action on board ship. But he was not, as we might expect, a doctor or a medical student acting as ship's surgeon. He was officially designated purser, but whaling ships had no use for a purser and this was only a device for signing on someone who amounted to a passenger. What arrangement had been made with the owners and with Captain William Adams is not recorded, but he probably paid handsomely for the privilege of seeking arctic adventurers on board a Davis Strait whaler.

Surviving relatives remember Livingstone-Learmonth as something of a rebel, loath to accept the dictums of a structured society and class, always restless to move on to new experiences and places. Born of prosperous Scottish parents near Melbourne, Australia, in 1861, he had received his education (with some reluctance it is said) at Edinburgh and London, then spent two years in the merchant marine, four in New Zealand managing his father's interests, a short period of residence in the United

States to study ranching methods, brief sojourns in France and Spain, ostensibly to learn the languages, and a hunting and fishing expedition to Iceland in 1887. In the following year he had joined the Peterhead whaler *Eclipse* under Captain David Gray for a whaling voyage to the Greenland Sea. Now, a year later, at age twenty-eight, he was off to see the wonders of Davis Strait and Baffin Bay. He hoped to be able to perform the useful work of mapping parts of the arctic coastline but in this he was to be disappointed. Such meticulous work could not be fitted into the *Maud*'s whaling itinerary during the short season. One suspects that his disappointment was not extreme, however, for in fact his real objective was otherwise. As he explained in a letter to the zoologist Robert Southwell, "I went in the *Maud* as a passenger, for shooting purposes...."

During the voyage, he would take every opportunity to get down on the ice or off in a boat, to shoot at animals of all sorts. His "bag" during the trip amounted to more than 400 birds of five species, twenty-six walruses, nine seals, and four polar bears.

Left: Polar bear cubs were sometimes taken home for sale to zoos. (Public Archives Canada, C-88363)

Right: The Danish official or "governor" at Noursoak, Greenland, and his half-breed wife, whom Livingstone-Learmonth described as "one of the most repulsive looking women it has ever been my ill-fortune to see." (Public Archives Canada, C-88274)

Many other animals were lost through sinking or wounding; he estimated that they were able to get only half of the walruses shot, and their success at retrieving seals can not have been much better. Dead and wounded birds lay upon the ice or bobbing in the wake of the ship. The slaughter was not totally indiscriminate and not entirely without economic justification, however, for in these last, desperate decades of arctic whaling the men would render into oil the blubber of seals, walruses, narwhals, and even polar bears, to supplement the meagre supply of whales. Yet, one has to wonder about the motives of a man who pays for the opportunity of snuffing out the lives of helpless animals with high-powered rifles and shotguns, and who so clearly derives pleasure in doing it—a man who describes a harpooned whale as "a truly fine sight," and who writes: "Walrus hunting is a grand sport... to see these huge brutes roll

over is a reward worth any discomfort to a healthy man"; and (after firing at a swimming bear from a whaleboat) "I shattered his skull with an express bullet. It was a grand chance for a photograph." Shooting eider ducks in flight pleased him immensely: "the splash of these heavy birds as they drop with a heavy thud in the water, is a cheerful sound, and dear to the heart of a sportsman."

Ironically, this same man was shocked at the sight of an Eskimo man crushing the skulls of seabirds with his teeth. "The man could have had no object in destroying those birds, except for the savage, innate, love of taking life," he wrote. Who, then, was calling whom "savage"?

Having made a voyage or two to the Arctic and asked a lot of questions, Livingstone-Learmonth fancied himself somewhat of an authority on the northern seas and an expert hunter of arctic animals. His assertive self-confidence amused veteran whalemen. Robert Gray, mate of the *Eclipse* in 1888 wrote, "Learmonth was one of those globe trotting individuals who knows everything and who has been everywhere. He knew how to shoot seals and bears before he had seen any of them." Despite Gray's advice, and that of his father David Gray (an arctic whaling captain for forty years), Livingstone-Learmonth shot at polar bears and seals from long range and aimed at their chests or shoulders; the result was often a wounded, enraged animal.

On one occasion in 1888, off the east coast of Greenland, the two Grays had watched from the *Eclipse* as Livingstone-Learmonth went after a polar bear on the pack ice. They delighted in describing the event later. Along with the captain's dog, Bob, Learmonth was transported to the ice by a whaleboat crew. Bob promptly leaped out of the boat, took after the bear, and brought him to a stand. Meanwhile, the Great Hunter, attired in an "ingenious attempt of some Bond Street tailor in the way of an arctic outfit," and "beladen as usual with rifles and cartridges," was "floundering waist-deep through the snow." Before he had covered a hundred feet he appeared to be "overcome...partly owing to excitement and partly to want of breath." The bear chose this moment to charge. Learmonth hurriedly fired but his shot either missed or merely wounded the animal, and it looked as if the game was up. But fortunately Bob made a timely attack on the bear's heels, giving Learmonth time to load and get away a second shot, which proved fatal. The hero of the story, in the eyes of David and Robert Gray, was the dog Bob.

The *Maud* sailed to Davis Strait in the summer of 1889, one of only half a dozen whalers in the region. There was no difficulty in reaching Disko by the end of April, but beyond this point ice made further progress agonizingly slow. It took a month to get up to Upernavik and another month to cross to Pond Inlet. Then, attempting to get westward, Captain Adams found Lancaster Sound completely and unexpectedly blocked by ice. But if the *Maud* and another Dundee whaler the *Nova Zembla* were brought to a halt, so were some migrating whales. In an exciting chase on the ninth of July, Adams permitted

Livingstone-Learmonth to pull an oar in one of the whaleboats and to try his hand at lancing the wounded animal. It was a cow whale attempting to protect her calf, and she fought furiously for over an hour, receiving three gun harpoons, four "rockets" (explosive projectiles fired from the harpoon gun), and a number of lance thrusts before succumbing. The men began flensing at ten in the evening and finished at four in the morning, obtaining approximately seventeen tons of oil (still in the blubber) and seventeen hundredweight of whalebone – a catch worth about £1500.

We take up the narrative of Livingstone-Learmonth two days later, on 11 July, near the mouth of Admiralty Inlet.

At 6 A.M. we heard the welcome cry of "There a fish," and two boats went away. The "fish" was a mother whale with a sucker accompanying her. Unfortunately they were scared and made good their escape. Our other boats had meanwhile been lowered and were lying at the floe edge ready for action. At 8.25 A.M. John McDonald seeing a "fish" lying head on to him, let his boat run down before a light

Left: Livingstone-Learmonth and two Greenlandic girls in front of hut at Upernavik, 1889. (Public Archives Canada, C-88250)

Top right: Title page of the journal of Matthew Campbell, written aboard the Nova Zembla *in 1884. (Dundee Museums)*

Bottom right: Maud *and* Nova Zembla *moored to the fast ice at Upernavik. (Public Archives Canada, C-88262)*

breeze, and fired his harpoon into the fish's "crown," as the rising pyramid shaped head of the black [bowhead] whale is called. Cheerfully we "called a fall," which means that all hands cheered lustily, and anxiously watched for the whale to reappear at the surface.

Now it may be as well to restate the exceptional ice conditions, which brought about an incident in whaling, which though not without a parallel, is sufficiently rare to make it a noteworthy occurrence. The ice was still "fast"—that is undetached from the shore—and stretching right across the Sound, from long. 82°w to the westward, owing probably to a prevalence of easterly gales. This fast ice acted as a barrier to the whales which were coming up the Sound to do their annual round, down Prince Regent Inlet, as far as Committee Bay, out of which they come in the end of winter, through Fury and Hecla Strait, out into

Left: One of the whaleboat crews of the Maud, *1889. (Public Archives Canada, C-88318)*

Right: Obtaining fresh water from streams in May 1889. Icebergs were another source, but were not always available. (Public Archives Canada, C-88278)

Davis Strait in the early spring. Thus the fish being unable to find a hole in which to breath to the westward of our longtitude were forced to await the opening of the ice later on in the year.

We left our harpooner "fast" in a whale, which proved to be a really plucky animal, as when struck it made a bold rush to the westward, under the solid ice. The struggle for life was a brave one, the means desperate, and the results fatal! With astonishing velocity, the whale ran out nearly three boats' lines (about two miles), and we were exceedingly anxious as with only one harpoon fast in a bad portion of the whale's body for holding the weight of these lines on the harpoon caused grave fears that the harpoon might draw out. After the whale had been away for one hour and forty minutes, the line ceased to run over the bows of the boat, and we knew that either the whale was drowned, or that the harpoon had drawn. With almost bated breath we took the lines to the steam winch when to our joy the heavy

strain assured us of the presence of the whale at the end of the line. For four long, weary hours we watched the lines slowly crawling in over the bows. There stood Captain Adams, and his face was a study. About £1200 hung on to the end of the line, and one sudden jerk, or careless management of the winch might have lost him this large sum of money. At last at 2.10 P.M. we hove the fish to the surface dead. It had proceeded so far under the floe as to be unable to return and thus had been drowned. Strange enough no sooner had we made the fish fast to the ship, and cleared the line from its tail and fins, round which it had become twisted than the harpoon fell out! How the men cheered when the huge animal made its appearance on the surface. It put our jovial skipper into high spirits, as even these two whales paid all the expenses of the voyage and left a handsome profit. This whale, which had bone measuring nine feet five inches, was calculated to yield thirteen hundredweight of bone, and fourteen tons of

blubber. I may here state that the present value of whalebone is about £1650 per ton. The whale measured forty-six feet in length, and was a medium sized male – or "bull whale" as the males are termed. Then began that most tedious operation of whaling – flensing. Right glad were we when the last of the blubber came on board. We flensed this fish in four hours.

While engaged in hauling up our whale we saw a number of Eskimo some walking over the ice, and a few coming towards us in kayaks. They proved to be some of that tribe which inhabit the land round Navy Board Inlet, and they had walked some thirty miles, over rough ground and ice to the ship. Poor things, they were very tired and the children cried a good deal, while the adults were in an excited state of exultation on finding that we were busily engaged in hauling up a whale. The women went on to the ice and prayed vigourously and with many gestures for the capture of the animal, while the men, with perhaps less

Left: Pond Inlet family and komatik, *1889. (Public Archives Canada, C-88389)*

Right: Fur-clad Eskimo boy at Pond Inlet, 1889. (Public Archives Canada, C-88361)

Far right: A woman and child at Pond Inlet, 1889. (Public Archives Canada, C-88388)

faith but greater belief in their deity helping those who helped themselves, hauled on the whale lines with a good will. Their delight on seeing the whale come to the surface dead was unbounded, and when our men gave the customary three cheers, they vociferated wildly. The captain had a large bucket of coffee made for them and gave them some broken biscuits, which they have learned to eat during the last two or three years. But they evidently did not care for the biscuits much for I saw many of them go down on the main deck and pick up pieces of the whale's black skin or blubber from among the men's feet, with all the attendant filth and mess incidental to flensing, and eat the very unpalatable looking morsels with evident relish. It was not a pleasant sight. Then as the "mollie" swarmed round the boats which attend to the harpooners engaged on the body of the whale, I saw one of the natives take up a position in the stern of one of the boats, and catch the birds with his hand, killing them by biting their heads. The fellow had accumulated quite a heap of them before he

was noticed, but this being a mortal sin according to a whaler's ideas, the man was soon routed out of the boat, and the poor birds were left to feed in peace. The man could have had no object in destroying these birds, except for the savage innate love of taking life, as the decks of the ship were lumbered up with blubber, which forms their natural food....

When on whaling ground our meals are taken at the most irregular hours, and we find ourselves breakfasting at two o'clock in the afternoon, or taking dinner at three in the morning, while we sleep when we can. But the constant tension of excitement enables us to do with very little sleep....

12 JULY

At 5 A.M., when most of us were sleeping the sleep of the just, we heard "a fall" called on deck. No sleep after that! Hurriedly rushing up we found that the indefatigable John Taylor, second mate, and best of harpooners, had got fast in a large whale a great distance away. Hurriedly we bundled all the Eskimo off the ship, to their great disappoint-ment, and left them standing—a cold and forlorn looking group on the ice. We then steamed out to John, as there were only two boats at the whale, and they might need more line. What a dance that whale led us. John had put both gun and hand harpoon into her—a rare feat. The whale dragged the boats about ten or fifteen miles. Then we got in two more harpoons, and the fish dragging the first fast boat against a loose floe and continuing her wild career, the boat came to the end of its lines, and being unable to pay out more, the lines were carried away. However with two other fast boats we felt pretty confident. The loose boats (boats which have not fired in a harpoon) then began a ceaseless persecution and although no one seemed to know exactly how many, I think about 6 or 7 rockets or bombs were fired into the whale until the internal arrangements of the poor animal must have been sadly disordered. Finally she rolled over on her back dead, amid thankful cheers, as we had occupied six hours killing her. The whale was a large female and had a young sucker with her, which I am glad to say escaped, as the idea of killing a very young animal, of little commercial value at that age, is repulsive. This fish measured fifty-two feet six inches in extreme length. Its tail was twenty-two feet six inches broad, and the length of the whalebone was eleven feet, which should give about eighteen hundredweight; while the blubber was estimated to yield sixteen or seventeen tons of oil. This was a great catch. In the Dundee whaling ships, the harpooner receives ten shillings and six pence and each of his boat's crew two and six, but if a harpooner puts in both gun and hand harpoon, as in this case, he receives £1-1 [one pound, one shilling], and each of this boat's crew five shillings. The *Nova Zembla* also killed a whale on this day.

It was blowing a strong gale when we finished killing our whale, and we had to tow her into smooth water. What a glorious uncertainty there is about whaling. Only seven days ago we were almost "clean," with only about fourteen tons of oil on board, and every prospect of having to stay out late in the autumn. Now we have about sixty tons and a handsome profit....

13 JULY

Flensing operations occupied us seven and a half hours, and were much prolonged owing to a gale of wind, fog, and short choppy sea. The fog lifted slightly, and we found that we were close in to the south shore of the Sound and had to make sail, and steam at full speed, to keep the ship off the rocks. It is most unpleasant to feel the ship making the usual rolling and heaving motions, as however strong the wind one does not anticipate a sea in these arctic sounds, where the loose and drifting ice usually form an efficient breakwater. There is however such a very large North Water this year in Baffin Bay that with an easterly wind the sea has space to acquire a good send....

16 JULY

At last the sun broke through the clouds, and right glad were we to see it after the fog, rain, snow and tempest of the last few days. Some of the Navy Board Inlet Eskimo came off in their kayaks in the forenoon, and stayed all day, sleeping on board during the night, or loafing about the deck, which appears to amount to the same thing with them. The Navy Board Inlet Eskimo have been guests of the Admiralty Inlet natives since the time when we so unceremoniously turned them all on to the ice on the twelfth, so these tribes live in harmony with one another. We took two of the kayaks on deck and I obtained their measurements, which were: length 20 ft. 6 in.; breadth 2 ft. 2 in.; depth 9 inches. The kayaks of these savages are much larger and very much more clumsily made than the kayaks of the civilized Eskimo on the coast of Greenland, which are patterns of neatness....

17 JULY

A small whale rose close to the *Nova Zembla* and boats from both ships went in pursuit. One of the *Nova Zembla*'s boats struck it and they eventually killed it. Poor fellows, this will help them a bit as they have been in bad luck. Each ship has now killed three whales, but by good fortune all of ours are large, while those killed by the *Nova Zembla* are small. We had a capital view of the whole hunt, and really when the ship is close you can see all the intricacies of a whale hunt much better from the ship than when you are in the boats.

Young ice began to form now in the water, when the sun was at its lowest altitude, or as we should say here – "at night." I shot about forty looms for the pot.

19 JULY

A number of Eskimo came off. One old fellow bartered to me his "kaiu," or tool with which they shape their ivory harpoons. He also bartered his "ichgagh" [igak], or spectacles, which are ingeniously made articles of wood with narrow slits in them to see through. These modify the glare of the sun on the glistening surface of the snow. Nearly all the old Eskimo have inflamed eyes, which inflammation is aggravated by dirt and their disinclination to apply a cold water cure. One begged me for a remedy so I supplied him with a small [jar?] of vaseline – an article of universal utility. In the afternoon two harpooners and I stalked a patch of walrus over the floe, and killed one, but it slipped off and sank. This was disappointing so one of the harpooners and I set off alone over the pack ice to try our luck with another patch. We jumped from piece to piece and had a mile of about as rough stalking as I ever wish to do. We were very wet, but at last arrived within sixty yards of the walrus, when not seeing any closer shelter I fired, killing one stone dead. The harpooner badly wounded another. They were lying on a small piece of ice and to our horror we saw the dead one slowly

slipping off and before we could reach it, it glided into the water and sank. It was cruelly tantalizing after our long and wearisome stalk. There was nothing for it but to make for the boat with as cheerful faces as possible. The harpooner in jumping from one piece of ice to another missed his footing and fell into the water up to his shoulders, but fortunately retained his hold of the ice. Poor fellow it was a chilly bath and when freezing hard a dip of that kind is not unattended with danger. Returning to the ship I fired a shot at a narwhal, which was lying on the surface of the water. The ball passed through its spine, paralysing it, but it sank before we could reach it. We made several attempts to harpoon narwhal but they were fruitless as the animals were very wild.

20 JULY

A number of natives came off in the forenoon, and a party of us went off in a boat to see the "toopiks" or tents at Navy Board Inlet, and also bring off some articles of barter. We landed on the edge of the fast ice, but found it much broken close in shore, so that we had much trouble to reach the land jumping from one piece to another. Nearly all of us contrived to slip in, and I had a good bath up to my middle to the intense amusement of the Eskimo who laughed heartily – and certainly the laugh was all on their side.

The toopiks are the usual miserable skin habitations, with the nameless odours inside. We got several narwhal horns and a few poor bearskins from the natives. While we were at the tents, the ship, which had been following the boat up, came round a point of the land in full view of the tents. The savages howled with delight. The whole tribe then set off for the ship which made fast to the ice, close to Adams Island. All went on board, and who shall describe the state of the decks when at last we got rid of them. A thick coating of black grease, dirt, and reindeer hair, of which skin their clothing is composed, and the hair of which is very brittle, completely covered the poop, and weeks of washing would not rid the ship of the pollution.

At last the moment which I had so much dreaded came, and casting off from the land ice the ship's head was turned to the eastward, the state of the ice precluding a further passage west. So ended my cherished hopes of seeing the capture of white whales (in Elwyn Bay); Sir John Franklin's first winter quarters at Beechey Island, where a tombstone is erected to his memory, and distant only about eighty miles; the winter quarters of M'Clintock at Leopold Island, about the same distance to the westward; and Fury Beach, where Sir John Ross wintered 1832–33 —and with these disappointments ended my chance of laying down unsurveyed coastline. All this is the more disappointing because the phenomenon of the ice lying fast across Lancaster Sound at this time of year has only twice been previously recorded during this century, and these are historic years in the annals of the Davis Strait whale fishing. In any ordinary year little difficulty is experienced in pushing much further west, and the whaling ships often go down into the Gulf of Boothia, where however many a good ship has been lost. ...

The *Maud* REMAINED A FEW MORE DAYS in the vicinity of Pond Inlet but without any further success in the whaling. As Livingstone-Learmonth learned, "the destructive pursuit to which the black whale has been subjected has so sadly thinned its numbers that but a stray member is taken here." No longer were flensed carcasses to be seen (and smelled) drifting in hundreds off the floe edge, as they had in the 1820s. The day of the bowhead whale had all but passed.

As for the ice, it was still the powerful and unpredictable adversary it had always been. Ice floes and bergs were still capable of crushing ships, both sailing vessels and steamers, whether their hulls were made of wood or of iron. Although the *Maud* and her crew had emerged unscathed they had more than once been reminded of the ways in which ice could influence ship movements and whaling success. They had been held up between Disko and the West Side for two months; they had found Lancaster Sound blocked by fast ice; and now, near Cape Hooper on the eve of departure homeward, they were beset in pack ice driven against the coast of Baffin Island by easterly winds. On 15 August Captain Adams succeeded in working the *Maud* out to the margin of the pack but not before ice had smashed the rudder and broken off a propeller blade. The steamer reverted to wind power and arrived back in Scotland on 11 September.

Robert Gray, chief officer of the Scottish whaler *Eclipse*, wrote somewhat facetiously in 1889 that Walter Livingstone-Learmonth was "already an authority on arctic affairs and intends to publish an account of the arctic regions soon." Learmonth did not realize this ambition, however, and regrettably so, because his observations of Baffin Bay scenery, settlements, inhabitants, and whaling operations are among the most interesting and most articulate of any. Furthermore his written record is supplemented by many of his own photographs, and while these are not the earliest taken in the Arctic they are nonetheless of special value because so little of traditional arctic whaling was ever recorded on film. Livingstone-Learmonth's contribution is therefore unique. Curiously, when his widow submitted his arctic journals to London publishers half a century after the voyages, they expressed no interest.

In a diary kept during his 1887 voyage to Iceland, Walter Livingstone-Learmonth wrote of some of the Englishmen present in the group: "Like ourselves [they] had come for the sake of an unconventional life in quest of exploration and sport." After the Davis Strait experience on the *Maud* his own quest took him to the open spaces of Argentina where he established a cattle ranch, married, and lived a relatively sedentary life until "retirement" at age forty-seven. He died in 1922.

The Maud *in a quiet moment. (Public*
Archives Canada, C-88396)

Whaleman in Eskimo kayak. *(Dundee Museums)*

Fifteen

The Inhabitants of the Arctic

I found quite a lot to interest me in a study of the ways of the natives. DAVID CARDNO, C. 1917

WHALEMEN HAD PLENTY of opportunity to meet Eskimos during the nineteenth and early twentieth centuries, but few of them had the time or inclination to record the characteristics of native life. Ships following the usual counter-clockwise circuit of Davis Strait and Baffin Bay and pursuing whales in successive localities as the season progressed would probably encounter Eskimo bands or families here and there, but only for brief periods. There might be time for some visiting on board ship or ashore at the native encampment and a chance to engage in casual barter, and the episode would likely find its way into the journal of a captain, mate, or surgeon. But there was seldom time for a close or continuous examination of the visitors. By necessity the observations of men such as John Wanless, Robert Goodsir, Stewart Lithgow, Thomas Macklin, and Walter Livingstone-Learmonth were limited to visible and obvious aspects of Eskimo appearance and material culture. On wintering ships and at shore whaling stations, however, where whalemen lived near Eskimos for several months or more, worked with them in whaling and other tasks, joined with them in games and entertainments, and sometimes accompanied them on hunting trips, it was possible to acquire some proficiency in their language and become familiar with their way of life. Yet Albert Whitehouse and many other whalemen who wintered close to Eskimo camps were more interested in setting down their own round of daily activities than in recording facets of native existence.

Some whalemen came to know the Eskimos as friends rather than strangers, as individual persons rather than groups of nameless curiosities, as partners in whaling and wintering rather than subsistence hunters concerned only with their own survival. Only a few, however, made systematic attempts to investigate, understand, and describe Eskimo culture. George Comer, an American whaling master with extensive experience in Davis Strait and Hudson Bay, was one such person. James Mutch, a Scots whaleman and station manager on Baffin Island, was another. Their knowledge, records, and collections were indispensable to the anthropologist Franz Boas, who after only one year in the Cumberland Sound region and no experience at all in Hudson Bay, was able to publish major works on the Eskimos of Baffin Island, Hudson Bay, and adjacent regions in 1888, 1901, and 1907.

A lesser-known observer of Eskimo culture was David Cardno

David Cardno, Aberdeen whaleman. (Mr. and Mrs. William Cruden)

from Aberdeen, whose brief, unpublished description of native life appears below. In 1866, at the age of thirteen, he stowed away on the whaler *Lord Saltoun*, sailing to Cumberland Sound, where his father was manager of whaling stations. The ship wintered at Niantelik, where the *Emma* had spent the winter of 1859–60 (chapter 10), and on one occasion young David, fleeing in anger and humiliation after a spanking from his father, moved in with an Eskimo family several miles away from the whaling harbour and lived with them for about three weeks – his initiation into the ways of the native people. A few years later, in 1868, he wintered in Cumberland Sound again on the *Lord Saltoun* and during the next thirty years he spent most of his time on sealing and whaling vessels destined for either the Greenland Sea or the Davis Strait region. In 1910, at the age of fifty-eight, he returned to Cumberland Sound as manager of the

two whaling and trading stations of Robert Kinnes & Sons of Dundee, living for a year at Kekerten and a year at Blacklead Island. He went out again in the same capacity in 1914, but on this occasion the annual supply ships failed to turn up in 1915 and 1916. Food ran out at both stations; the native hunters had to be released to fend for themselves; and Cardno subsisted on what he came to consider a monotonous diet of caribou, salmon, and birds' eggs – drinking hot water instead of tea and coffee. He passed the time reading, playing solitaire, and writing in his journal; there was time, at least, to set down a few details of Eskimo life that he had observed during nine winters in Cumberland Sound.

THE ESKIMO

The inhabitants of the Arctic are the Eskimo, also written Esquimaux, Usquemows, called by themselves Innuit (Innuk singular; Innooeet plural), or "the people." The word Usquemow is Indian, meaning raw flesh eaters, the English and Scotch anglicized it to Eskimo. The name "Husky" as applied to the native is…slang, a corruption of Eskimo perpetrated by men whose ears and tongues were untrained to the language – whalers who sometimes employed the tribesmen in their hunting and dubbed them with the first jargon name that came handy. It is still used in this sense in localities where Europeans are numerous, such as Alaska and Hudson Bay.

NATIVE TRADE

Even before the advent of Europeans and the trade they brought with them there was a certain amount of barter going on among the arctic folk themselves, occasioning not a little movement. More driftwood being found in some localities than in others (chiefly at a place called Tudjadjuak), the tribes came from everywhere to barter for it with those on the spot. Again, the soapstone and potstone of which their lamps and cooking utensils were made, is found in a few places only, such as Katwaq, Kekerten, and

The Blacklead Island whaling station (right),
Anglican mission (background), *and native
tents* (left foreground), *1903–04. (Public
Archives Canada, PA-53579)*

Nuarmaqdjuin, so that the natives came a long distance to dig or trade for that too. Pyrites for striking fire was also a valuable (if local) production, and flint for arrowhead-making.

THE BUILDING OF THE VILLAGE

The Eskimo are a wandering folk thus their dwellings must of necessity be capable of quick erection, demolition, and easy transport. The tribe lives in tents in the summer, moving from one camp to another as the hunters decide, but winter quarters are more permanent and the snow-built house, the igloo, takes the place of the sealskin tupik on a more lasting foundation. The Eskimo tent is a wholly different affair from an Indian wigwam or lodge. It consists of a penthouse-shaped framework of poles semi-circular [at?] the back, with overlapping stripes of curtains of dressed skin for the entrance in front. The whole thing carries a covering of skin firmly and beautifully stitched together. The back part of the tent used as the family sleeping place, is covered with skins of the large grey seal with the hair left on in order to insure some darkness during the long unbroken day of the arctic summer. The heavy hair also serves to throw off the rain in wet weather. But the front portion of the dwelling has a roofing of thinner membranes of the sealskins pared from the entire pelt when fresh and moist. These membranes are first stretched upon frames and dried prior to being sewn together, when they become almost transparent, so that there is plenty of light in the rest of the tent. They are so beautifully and neatly stitched as to be practically waterproof.

BUILDING AN IGLOO FOR WINTER

The main considerations the Eskimo has to bear in mind in building his snow house are that it will have to be kept in repair and that it must be adequately lighted and warmed. This means labour and oil. So for his own sake the dwelling is planned on as small a scale as possible. It varies

in nothing but in this point of size from all the rest of the village. The hunter having found his site next takes his sealing spear, a long twelve-inch knife, and a saw, and begins piercing the snow in every direction, his object being to find a spot where it is deep snow, and so closely packed and hardened by the wind that it can be cut out into great blocks for building. Otherwise his "bricks" would be too brittle or too friable for the purpose. Should no such patch be near at hand, the builder calls all hands. They start trampling and packing the snow with their feet while the old men, the women, and boys constantly bring up fresh supplies and throw it in to be stamped firm. Having thus prepared what he considers enough material for his purpose, the Eskimo commences to saw out huge rectangular blocks of this solidified snow mass, each one of which taxes his utmost strength to lift. He begins his house by building a ring of them, a larger or small ring as the case may be, fitted and jointed together with the utmost nicety by means of his knife. A second tier is added to this ring, the builder working from the inside and the blocks

Left, right (and following pages): Drawings from David Cardno's journal. (Mr. and Mrs. William Cruden)

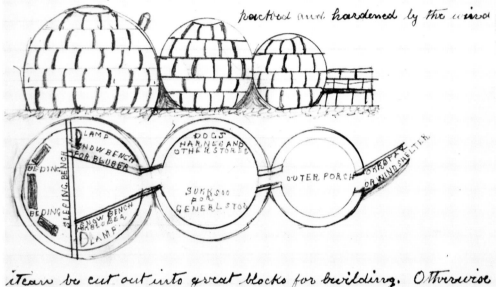

being brought up by his assistants. As soon as this is well and truly laid, he trims the upper surface to a slope and continues building, but in a spiral now and slightly sloping inwards until he has reached the top of what has grown to be a dome roof. A key block is deftly fitted in to complete and close it, and the shell of the dwelling is complete. The finishing touch is the window – a slab of fresh water ice. He cuts a block of snow out and then inserts the ice and then it is glazed, the fashion of the Arctic.

THE SEALING GROUNDS

The hunters start out early in the morning after a hasty meal of raw flesh and a drink of water, accompanied by their sons, and the dogs – four or five in number – harnessed to a light sled loaded with canoe, spear, harpoon and line, or whatever implements may be needed for the proposed chore. The teams start out in a fine tear, urged by shouting and the cracking of whips, and off they all race – sleds and dogs together – to the sealing grounds out on the frozen sea (or inland for the caribou). The stars serve as a compass or in thick weather the [word missing] will be sufficient guide.

...Directly he has detected the locality of a seal's nursing cavern under his feet [between the snow and the ice surface, close to an access hole through the ice], either by the presence of a slight depression in the snow, or by the pointing of the dog, he arms himself with a "nixir" or hook on the end of a long shaft and gathering himself together

off the baby's retreat into the sea beneath.
the debris of the cavern for the imprisoned
hooks it out and kills it with one blow on

SNOW   SNOW

SEAL
YOUNG

ICE   ICE

MOTHER SEAL

WATER   WATER

Womanhood in the arctic

makes a tremendous jump into the air coming down with all his might and force upon the spot. He jumps again and again until at last the snow caves in and blocks the hole below cutting off the baby's retreat into the sea beneath. Then he prods among the debris of the cavern for the imprisoned creature, locates it, hooks it out, and kills it with a blow on the head.

WOMANHOOD IN THE ARCTIC

In the meantime the women left in the village on shore have been far from idle. As soon as the husband has gone off for the dog the wife sets about her domestic affairs. First she rolls up the bedding and tidies the sleeping bench. The next job to sweep hoar frost from the window and the cupola to prevent the dripping of any moisture, and then to sweep up the floor—littered, likely enough, with the remains of a good feed overnight. These duties are performed with a brush made of the outspread wings of a duck or a raven; it might almost be called a double-bladed brush. The backs are sewn together and the upper

bones form the handle. Such a contrivance is a very handy affair altogether and will last a long time. The next task is to prepare a quantity of blubber for oil. This is pulped with a bone hammer or "koutak" and the fuel so obtained is suspended over the shallow lamps in such a way as to empty into them and keep them supplied. A new wick is fashioned from dried moss and cotton plant trimmed upon the lamps. Next comes the stew for supper. The Eskimos have only one way of cooking meat and that is stewing it in the stone pots already described. These are partly filled with sea water for the sake of salt, and fresh water added till it is to [one's] taste also seals' blood, and then comes meat. The whole thing hangs simmering over the lamps all day, and by the time the men come back at night a [boiling?] hot meal is ready—rich, nourishing and as tender as a sharp-set hunter could desire.

CLOTHING

...The Eskimo woman values none of her possessions more than the "ooloo" a short-handled knife shaped like a half-moon turf-cutter, chiefly used for paring off the inner membrane of the stout sealskin for the lighter hangings of the summer tent, but of universal utility. With it she cuts out her garments or dismembers a seal. In addition to this she has steel or heavy ivory needles and a thimble. The Eskimo had no woven garments or European clothes until they came in contact with the white and perhaps unfortunately acquired the beginnings of a civilisation alien to the natural evolution and necessities of their lives. Their own natives' dress consists entirely of deerskins for the winter use and sealskin for summer. Both sets are warmly lined with fur. The deerskins employed as clothing are the summer and autumn hides, those employed in the winter are reserved for the "kaksak" or sleeping blankets. The men's and women's tunics are lined with either fawn skins or the summer skins with the hair on. No under-clothing is required, fur always being worn next the skin.

The man's jacket is looser in shape than the women's and the hood ("nessak") fits closely round the face. The women's garment is quite different. It has shorter baggy sleeves, is large and roomy at the back, fitting, however, tightly to the waist, it has a hood big enough for two heads, a short stomacher-like apron about twelve inches long in front and a lengthy tail reaching to the heels behind. The Eskimo women carry their babies on their backs in this queer jacket. The child has no clothing on it [but] keeps admirably warm next the fur-clad mother. Its feet rest on her waist line and its head peers from out the capacious hood over her shoulders....

Prior to the advent of Europeans to the Arctic, fringes of deer skin were the most popular form of ornament for clothing. But today the Eskimos women are passionately fond of elaborate beadwork. The beads are of European manufacture but the designs in which they are applied are native. The favourite beads are small and brightly coloured. The native seamstress will also sew two or three coins down the front of the inside jacket and down the tail of the dress or even the bowls of a few spoons; they clink as they walk and greatly delight their wearers.

THE KAYAK

Perhaps the next important business of the Eskimo woman, after cooking and making the clothes is the preparation of skins for the two types of boat in use on the coast. This entails considerable labour and skill. The men are responsible for the framework. The kayak (a creation as truly national to these intrepid coasters as the snowshoe may be to the Indian, the ski to the Norwegian, and the alpenstock to the Swiss mountaineer) is a covered canoe, graceful as a fish, for use at sea. It can be handled in the roughest weather. It consists of a light framework, formerly of whalebone but now generally of driftwood, fastened together with thongs of seal skins. It is from fourteen to twenty feet in length, strong and elastic to a degree, and

entirely covered with skins, almost resembling a torpedo in shape with tapering extremities. There is a small circular opening amidships where the kayaker sits, fitting closely round his body. In rough weather he wears a waterproof jacket (of seal gut) the hood fitting snugly round his face and sleeves to his wrists. The lower edge of this comes over the opening in the canoe and is laced round it so that man and craft are fairly one.

The sealskins for these canoes are bleached. Either they are scalded or tied up in bundles and hung up in a warm atmosphere to ferment. This process is allowed to go on for a week or two, until the stench becomes unbearable. When taken down and shaved with the "ooloo" the black epidermis comes away with the hair, leaving the skins beautifully white. The inner membranes are left intact. The next step is to stitch the skins together. Bleached hides may be made to alternate with unbleached ones by way of ornament: or the entire covering may be black or brown.

When [the] canoe is fully fitted out...[the hunter]... carries a three-pronged bird spear on the left hand side in front of him. On the right is his sealing spear, and between the two is a small round tray for the coiled sealline fixed to the detachable spearhead. Behind him on the left is the "nixir" or hook, on [the] right a heavy harpoon for striking walrus or the large creatures he may encounter. Between the two and immediately behind him [is] an inflated sealskin with the end of his sealing line attached. Thus equipped the canoe is complete, a thing of pride to its owner.

THE UMIAK OR SKIN BOAT

The umiak is a very different craft and serves the Eskimo family as a sort of general pantechnicon and removing van. It consists of a large sturdy framework of wood covered with the skins of the big ground seal, which are dressed into a thick, tough, leather. It is really an open sailing boat, capable of carrying perhaps six families and a huge and miscellaneous cargo. It has a square stern...and a stumpy mast set well forward in the bows. The large squaresail used to be made in earlier days [of] skin stitched together or of the intestines of seals blown out and dried, then split open, the long broad strips alternating with narrow strips of [the] same material to ensure equal stretching and shrinking. Nowadays the natives provide themselves with sailcloth from the trading post. The umiak is an unhandy thing to manage; as she has only two very clumsy oars to pull with it [takes] them a long time to get [anywhere?].

ESKIMO DOGS

The value to the native of a good team of about five to eight dogs is equivalent to that of a kayak, or a sled, or a reliable gun. To assess it in terms of money would have no significance in a land where utility and necessity alone determine the scale. The breed is part or half wolf. In build the true Eskimo dog is well-formed, almost slim about the hind-quarters compared with the rest of his body, the broad and sturdy chest, the strong neck and heavy jaws. His hair is very thick, grey or tawny in colour, and his tail immensely bushy, always carried erectly curving over the back. He is a different creature to the Samoyed and the Kentucky wolf hound: but probably there is very little to distinguish him from the famous Alaskan "husky dog" of so much literary fame and the dog of the Labrador.

THE "KUMMATIK"

The "kummatik" or long travelling sled is double the length of the foregoing [hunting sled?] and heavier in proportion. Otherwise its construction is the same. It requires a team of from twelve to eighteen dogs whereas five are sufficient for the hunting sled. The loading of a "kummatik" is a work of art. There is a place for everything, and everything has to go just so into its place.

The spears and [word missing] are stowed in the bottom of the sled in front by the driver. At the far end a piece of skin is laid down and upon this slab upon slab of blubber for the lamp is piled up and the lamp set atop of the lot, bottom up because of the grease and dirt. Then the meat for the journey is put on board – frozen deer hams, and frozen seals entire, enough for the whole party until they fetch up at the next tribe's camping ground. The meat is of course uncooked since a minimum of raw meat gives a maximum of heat and strength. Hence the Eskimo prefer their rations raw when there is work to be done. The cooked stew of [an evening?] is a mere luxury meal. A skin is thrown over the heap of provisions to prevent the travellers' clothing being soiled by it. Over it all are piled the rolled-up sleeping blankets and the "harsate", or deerskin rugs, for mattresses. Knives, axes, and lines hang on the horns behind. The driver's seat in front is a box containing small tools, flint and steel. The whole lot is securely lashed down to the crossbars of the sledge. The ancient form of sled as described by the oldest hunters in the past when whales were plentiful and the whalebone of great value to the Eskimos: strips of whalebone were stitched together with whalebone throngs and a flat sled formed. It was very strong and less liable to sink in the snow.

TRIBAL LIFE

In their family or tribal life the natives carry out a very smooth-running sort of communism, the chief tenets of which are rigidly enforced peaceableness, open hospitality to the stranger, and a sharing of food and the necessaries of precarious existence among each other. Tribal government is wholly patriarchal in character. The "angakooeet" or chief conjurors – a class of men apart – hold the first place in public esteem and common council. After them the village is ruled by the successful hunters who foregather with the former and with the aged at the far end. a piece of skin is laid down and

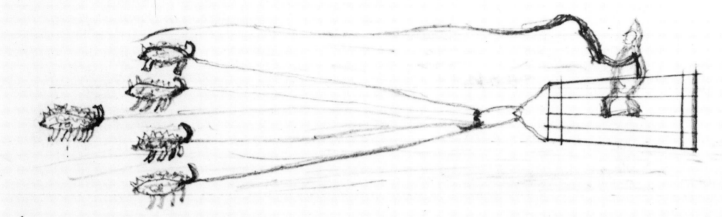

The Ancient Form of Sled as described by the oldes

and experienced when it is a question of deciding where to go, and what to do about the hunting, or change of encampment, or treatment of a delinquent.

The Eskimo community in sanitation or in sex matters has few reticences. This may be another way of saying it has no pruderies. The native attaches no more importance to the functions of sex than to those of eating and drinking or sleeping. It would of course be easier to attribute complete insouciance in these respects to the native mind if, instead of trapping some of them out with rather elaborate ceremonial, it kept all much on a level in most instances of insistence. However, a hygenic motive, conscious or unconscious, lies behind them. Although the people live under very crude conditions crowded together in the igloo, with [no] privacy or special quarters for women, they are not without a sense of the fitness of things or some idea of personal modesty. It is the height of ill-breeding to stare, for instance, at anyone whilst dressing or undressing. Like the Indians and like most other uncivilised people the Eskimo marry early, sometimes indeed at the age of twelve years. Unions are arranged by the mothers and grandmothers. A woman with a marriageable daughter is fully alive to the advantage of seeing a good hunter attach himself to the domestic circle.

CHILDBIRTH

A girl will be attended in childbirth with her first baby but not after that. The expectant Eskimo mother has to be alone (except on the first occasion) in a little house set apart for her, and without assistance. After the baby is born the baby is never washed but rubbed down with a soft fur or bird skin and put straight away, stark naked, into the capacious hood of its mother's tunic. The woman must, however, never eat alone during this time, lest a "tougak" with three fingers steal her food and bring evil upon the child. She must pay no visits until she has quite recovered in the space of a full month, and only then if she has a new suit of clothes.

DEATH AND BURIAL

These people fear death and the dying. Just before a man dies he is dragged outside the house or tent so that his spirit may not haunt it. No dwelling where a death has taken place is ever reoccupied. Should anyone chance to die inside all the possessions are held to be polluted and must be cast away. A corpse is sewn in the deceased's accustomed sleeping blanket, placed on a hand sledge, and hauled away to the chosen place of burial, followed by the members of the family and the relatives, it is laid on the bare [ground] (the ground being frozen hard as iron, grave digging is out of the question) and huge stones are piled around and upon it like a cairn. In the case of a man his weapon, drinking cup, and knife – or these things in miniature – are placed beside him, his sled or a small model of it nearby, and he is buried with a little sort of doll representing a woman. In the case of a female her needles, knife, cup, and a man doll, are laid beside her. Food is deposited on a flat rock near the pile and the mourners sit down to eat a farewell meal with the spirit of the dead. Then they march in single file seven times round the cairn following the direction of the sun, that is, from east to west, chanting directions to the departed.

CONJUROR OR ''ANGAKOK''

For phantasy the writer is aware of course that the beliefs of the Eskimos are paralleled by those of many other uncivilised people. ... Briefly put, the Eskimo religion consists in the belief in a multiplicity of spirits, good and bad, and in one Supreme Spirit of whom no fear is felt because he has no evil intention towards man. The conjuration and propitiation of the evil spirits is the constant business of the conjuring class, although everyone has some degree of power to deal with them. Man was made indeed by the Great Supreme Spirit and his name was given Akkduksa, and women, one Omaneetok, was fashioned from his left-hand floating rib. The Eskimo very highly esteem their own race, but hold Europeans in considerable

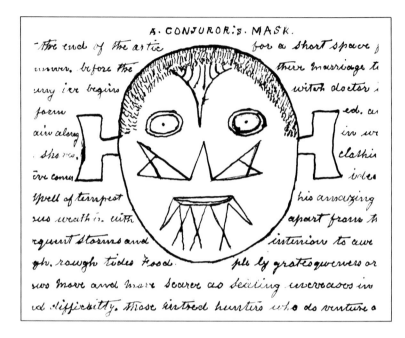

A. CONJUROR.s. MASK.

contempt. They have an unpleasant legend of a woman and a dog being cast away together in a boat or on a floe, by way of accounting for the origin of the whites. Man's spirit, like the spirit of everything else, is immortal and destined to a future life in bliss, in the region where the Great Spirit presides over a happy community of very properous Eskimos such as has already been described. Those who die on the hunt go to this heaven, also women in childbirth, and those who die a violent death by any sort of accident. The road to this Eskimo heaven is beset by many obstacles and pitfalls. It is haunted by savage animals who lie in wait to attack, maim, and kill the wayfarers upon it. Legend has it that at the end of this road, at the rim of this world which is the gate to the next, two huge rocks are set confronting each other across the narrow path. They sway ominously and often crash together, so that the soul seeking heaven has to run the risk of being caught and crushed between them as he endeavours to get through.

...At the end of the arctic summer before the young ice begins to form again along the shores there comes a spell of tempestuous weather, with frequent storms and high rough tides. Food grows more and more scarce as sealing increases in risk and difficulty. Those intrepid hunters who do venture out return empty-handed day after day and it grows high time for something to be done. The goddess Sedna is supposed to be causing these storms and all this dirty weather at sea to prevent her animals being...killed. And so a conjuration has to be performed to liberate the seals.

This is the occasion of the most elaborate festival in the Eskimo calendar. It begins by the conjurors in full dress calling the people altogether to dispense them for a short space from their marriage ties. Each witch doctor is masked and clad in woman's clothing. The idea of his amazing get-up, apart from the usual intention to awe the people by grotesqueness or hideousness, is to disguise the face and body, to efface (as it were) the well-known individual, to make the people lose sight of conjuror in the representation of a Great Power at work among them. His dress is partly that of a man and partly that of a woman, and he carries the usual implements used by both sexes. This is to bring the needs of either before the Great Power and to intercede for their respective needs.

THE NATIVE SURGEON

Eskimo flesh has wonderful healing power. The writer has seen the most fearful gashes quickly close and heal up without any precautions or dressing whatever. One case he certainly would have [expected?] a fatal termination. A hunter was repairing his implements, a small box of tools lying on the ground beside him. A large file without a handle happened to be sticking straight up out of the box. The man's foot slipped on the ice and he fell in a sitting posture straight upon the file. He sustained a deep punctured wound. It was merely bandaged with some dirty strips of soiled skin underclothing, and inflammation and intense suppuration prestly [i.e., promptly] set in. At no time did the wound receive any further attention but in

due course the hunter was about again as though nothing has happened.

Something however, must be said for the conjuror as an anatomist. By virtue of his calling and of his continual dealing with animals of all kinds he knows the positions of joints, muscles, ligaments, veins, and arteries, and can find any one of them. Some men have more aptitude in this respect than others and thus occasionally act as surgeons. A young woman who[se] name was Omenak, the daughter of one of the hunters, developed a large swelling in the groin. There was inflammation pointing to deep-seated pus in accumulation. A native surgeon was called in and after examination he pronounced for an immediate operation. He decided to lance the swelling. A time was arranged and by special request the writer was allowed to be present. The surgeon arrived. The patient was then prepared by her mother. She was laid flat upon the bench and the part to be operated was exposed. The surgeon wetting his fingers in his mouth proceeded to moisten and clean the skin. Their two assistants grasped Omenak by the legs. Her mother held her, and two brave helpers held her well down by the shoulders. The conjuror inserted the lancet simply by pressing on it and sawing it in backward and forwards until it had gone deep enough to reach the pus. Omenak squirmed considerably but her nurses had her well in hand. The contents of the swelling were expelled by repeated pressure and wiped away from time to time with a little bit of lemming skin. When this was finished the wound was covered by a piece of lemming skin licked by the operator's tongue and stuck over the cut. Two days afterwards the patient was walking about well, and jolly as ever she had been in her life.

SPORT AND HUNTING

They pull away joyously and hilariously on the great summer trip. As often as the wind will allow they hoist the sail made of seal intestine and one member of the crew takes up a station beside it with a water bucket to keep it constantly wet. Otherwise, it would dry and split in ribbons before the breeze. At the present day duck [light canvas] sails are used.

Every now and again as they coast along the islands they put in here and or there for fresh supplies of drinking water. At night they fetch some well-known point for an encampment. The boats are moored, heather and drift-wood collected, fires lit, kettles slung, and the evening stew set to simmer while they forage afield for the next day's provender. Then rolling themselves up in their blankets the travellers drop off to sleep right there on the ground under the shelter of whatever shelter can be found, to be up and away again before the sun is very high in the heavens.

The days pass very pleasantly. The scenery is grand, the weather clear and warm, the water gemmed with islands dark brown and green, as still as a mill-pond. The fleet of boats filled with men and women of the Arctic, with tattooed faces and guttural speech, reproduces a picture of prehistoric times. Many of these scenes of Eskimo life and enterprise are deserving of record by the best of artists, if to bring before us in the effete day of over-civilisation a vivid, still existent, picture of the very earliest adventures of the human race.

At length the head of the fiord is reached. The boats are hauled up on the land to be left there until they return from the hunt. They are all unloaded and turned bottom-up, and some stone put on to them for weight to keep them from blowing away. The personal treasures of the women are also put away in some safe cavity among the rocks and left there. Then the loads are carefully apportioned all round and made up in bundles according to the strength of the carriers. The men bear the weapons and ammunition and travel light in order to go on ahead and secure game on the trail. Children are lightly loaded and the old people carry little or nothing, so that the bulk of the heavy transport falls on the able-bodied women of the tribe. Each one toils along under tent poles and

coverings, piles of skins and meat, and the baby of the family into the bargain. The whole staggering load is hoisted onto the woman's back and secured by lashing round the waist and a broad leather band round the forehead. She is almost wholly eclipsed by the enormous burden.

There are contests with the bow and arrow. Poles are fixed in the ground with skins [suspended?] from them to represent deer and seals. The vital spot of course is the natives' idea of the bull's eye. The spear-throwing competition calls for a high degree of skill. From the top of a fixed pole a line is carried to the earth having an ivory ring tied in it half way down. This ring is carefully concealed by fringes of hide and the spear throwers, stationed at a recognised distance away, have to cast their weapons deftly through it. The attempt demands the greatest accuracy of vision and training of the hands. The contests are very keen and great éclat awaits those who distinguish themselves. Their names become household words round the igloo lamps all during the succeeding winter, much as those of crack footballers become familiar to the sporting manhood of this country.

BEAR HUNTING

Bear hunting again is pursued by the Eskimo with no less zest than that of the seal or deer.... The bear is much respected by the Eskimo for his intelligence and cunning and his strength. Indeed they consider [bear] second only among the creatures of the wild to man himself. It is for this reason that they so often choose for their "tongak" – or guardian familiar – the spirit of a bear. One very curious belief about the animal is that the bear himself has a "tongak," quite distinct from his soul or "tarngnil," and that when this spirit requires a new commodity such as a new seal warp or line, which is represented by the black skin round [the] mouth of its [protégé?], this "tongak" causes the bear to fall in the hunter's way and be killed. The hunter spares the black skin and refrains from cutting it when flaying the carcass as an offering to the spirit. A further offering of the sort is made by transfixing various portions of the beast's body and entrails on a stake together with a man's implement such as a knife if the bear were a male, or a woman's implement such as a needle or skin scraper if it were a female, and exposing the gift for three days. At the end of that time it is thrown into the sea. In bear hunting the rule is for the skin to go to the first hunter who sights the prey (not necessarily the first to kill it). The best part of the body goes to him who deals the fatal blow. The arctic bear is not a hibernating animal for it is only the female who sleeps through the winter. [The male returns]...to the female only in the spring when she emerges from her hiding place, gaunt and hungry and accompanied by the cubs. He is always the safer creature to hunt at such a season, since the female is then thoroughly out of condition and very savage.

The wolves and foxes were trapped by the hunters in...[this] manner. A small igloo was built in the ground ice along the seashore where it would not be conspicious and a loaded gun fixed pointing to the entrance, which did not have space for anything but forward movement. A trail of meat led to the entrance, inside of which was a piece of meat...tied to a string. The other end of the string was attached to the trigger. The wolf entered, seized the meat, and shot himself.

A SEAGUL TRAP.

The skins of...[seagulls] are used for slippers which go over the fur stocking and inside the boot to prevent the cold striking through to the foot. The old hunters build a small igloo amid the rough ice of the seashore, leaving a hole in the top. Pieces of blubber are scattered on the top to attract the gulls, who alight by the side of the hole and are caught by the legs and dragged inside. The flesh is eaten.

WHALING EXERTED A POWERFUL INFLUENCE upon the native people of the region during the nineteenth century, particularly the inhabitants of the West Land, or Baffin Island.

On the East Land the impact of whaling was not so great, partly because the Greenland Eskimos had already experienced substantial cultural change through contact with European vessels and Danish colonists during the eighteenth century (and Norse settlers several centuries earlier), partly because the Danish authorities attempted to restrict economic and social relationships between foreigners and native Greenlanders, and partly because the English and Scottish whalers were normally transient along the Greenland coast, not inclined to stop long in any locality to pursue whales, winter, or establish shore stations. Their brief visits to Lievely (Godhavn), Upernavik, and other settlements were normally occasions for good-natured barter between sailors and natives for small items of a souvenir nature, opportunities for a little music and dancing on deck, pleasant interludes of fun and festivity to punctuate the serious business of whaling. A few visits resulted in friction and hostility between British and Danes, or Greenlanders. Whatever their nature, these sporadic contacts did not lead to systematic trade or employment.

At the top of Melville Bay the more primitive inhabitants of the Cape York area occasionally came off to ships. This was more of a no-man's-land, remote from direct Danish super-vision, but the whaling captains were never anxious to linger in this region for a trivial trade in skins or seal oil; their eyes were set on Pond Inlet and Lancaster Sound, where whales were known to be abundant.

On the West Side the Eskimos were totally unhindered (and unprotected) by colonial authority. When the whaling frontier first reached their shores about 1820 they were living in the traditional manner that had evolved through the centuries and had been modified only slightly by occasional contact with explorers and the acquisition of a few European implements, bits of metal, and pieces of wood. They were hungry for European goods, and were quite willing—even eager—to work for whatever they required, using their considerable skills as boatmen, hunters, and dog drivers. The potential for cultural modification through whaling contact was therefore very great. By the end of the whaling period substantial changes had taken place in Eskimo material culture. Among the hunters there were rifles, telescopes, sheath knives, jack-knives, hatchets, saws, drills, awls, steels, and files. The women cooked in metal pots and kettles, and used steel needles, cotton thread, and metal scissors in sewing. Luxury items such as accordions, fiddles, and sewing machines existed here and there. The most visible change was in clothing, not so much in the winter attire (for furs and skins were unsurpassed in lightness and warmth) but in the summer dress. Women had adopted woollen shawls and long skirts of cotton. Men wore cloth trousers, shirts, jackets, caps, and even waistcoats. Incongruous, "trendy" items such as pocket watches and chains, bowler hats, nautical caps, and sunglasses were sometimes in evidence.

Whaling masters and station managers could offer an impressive variety of manufactured goods, not only for trading purposes but also as payment in kind for services rendered by Eskimos; many of the articles could be enormously useful in routine domestic operations or in travelling and hunting. The Eskimos, on the other hand, could not call upon the productive resources of a factory system to provide a surplus of goods, let alone goods that would be appealing to representatives of an industrial society, but they did succeed in marketing certain items of clothing among the whalemen, in particular waterproof sealskin boots and mittens, and hooded parkas of caribou skin. These

were widely used by the whalemen, especially when wintering.

In 1852 Captain Quayle took the *McLellan* north from New London, Connecticut, to pick up the twelve men who a year earlier had volunteered to spend the winter ashore in Cumberland Sound. George Tyson, on his first arctic voyage, had been among them (chapter 12). Unfortunately the ship was crushed by ice in Melville Bay, but Captain Parker of the famous Hull whaler *Truelove* took Quayle and some of his men on board and headed to Cumberland Sound to complete the mission. As they approached Niantelik—everyone anxious to learn whether the men were still alive—they sighted an American whaleboat, and sailed over joyously to meet it. William Barron, then an apprentice on the *Truelove*, later recalled the moment: "When we first sighted the boat, all hands were on the tip-toe of expectation, but as it came alongside, everybody was sorely disappointed. Instead of the American crew, we found they were Esquimaux, in European clothing." The *Truelove* continued to Kingmiksoke but there was no sign of the men there. The next morning, however, while the whaleboats were out cruising, Barron and his mates saw another boat, apparently manned by natives, and went over to investigate. It turned out to contain some of the men from the American wintering party, and they were all dressed like Eskimos. Evidently the diffusion of material goods between the two cultures proceeded in both directions.

By the First World War and David Cardno's residence in Cumberland Sound the Davis Strait fishery had ended. Dundee, the last port remaining in the business, had sent five steam whalers out in 1911 but they had succeeded in securing only one whale. In 1913 Dundee dispatched no whalers. Commercial whaling was finished, not only in the Davis Strait region but in Hudson Bay and the Beaufort Sea as well.

For three years Cardno was the only white man among two or three hundred Eskimos at Kekerten. Across Cumberland Sound another man, James Law, eked out his own separate and lonely existence at the Blacklead Island station, near a few hundred Eskimos of that region. Two white men, anachronistic relics of a dead industry, cut off from their homeland, unaware of the World War until two years after its outbreak, forgotten by their countrymen, and ignored by the Eskimos. After almost a century of exploitation by European and American whaling vessels Cumberland Sound had almost reverted to its former solitude and to its original inhabitants, or so it seemed.

But the fingers of European exploitation were still hooked into the arctic regions. At half a dozen small stations in the Canadian Eastern Arctic (Kekerten and Blacklead among them) a few representatives of British and American companies continued to hire native hunters to secure what were loosely referred to as "scraps" – the hides of walrus, seals, caribou, and bear, the tusks of narwhal and walrus, oil from several species of sea mammals, and so on. A few small sailing ships visited the stations when they could, to collect the produce. Trading vessels occasionally plied the coastal waters, picking up what they could from the Eskimos. These operations were economically marginal, undertaken by companies without much capital, often depending on chartered vessels of dubious quality, with inexperienced or incompetent crews.

The European system of procuring wildlife resources from the Canadian Arctic was not ending, however, but simply changing. Whaling itself had evolved through the years from wide-ranging pelagic hunting at the outset, to extensive whaleboat cruising at particular localities, to wintering with floe edge whaling in spring, and, finally, to year-round shore stations employing native whalemen and hunters. From an initial dependence upon one quarry, the bowhead whale, the whaling industry had turned to a diversity of species and products as whales became scarce, finally relinquishing almost entirely the catching of whales and other sea mammals to native hunters. Large, well-capitalized companies had abandoned the field to small companies and individuals. The final stage of the economic transition was from scattered, unco-ordinated "free traders" operating on shoe-string budgets, to the integrated, monopolistic system of trading posts established through the entire Canadian Arctic by the Hudson's Bay Company, beginning in 1909.

Fundamental changes thus occurred at the end of the arctic whaling period, and these changes affected the native popula-tion in important ways. Once active partners in whaling and sea mammal hunting, the Eskimos of Baffin Island were now to become trappers of fur-bearing animals. Attracted during the nineteenth century into relatively few large settlements at which opportunities for trade and employment were best, and frequently employed during the winter to procure game for ships and stations, they were now to be dispersed over the land again in the quest for furs. Given the opportunity in the era of wintering whaleships to experience regular face-to-face contact with a large number and variety of Europeans and North Americans, they were now to limit their inter-cultural encounters to brief trading transactions with one or two whites at a trading post. The change from whaling to systematized trading would reduce the frequency of contact between Eskimo and whites, retard cultural exchange, and inhibit mutual understanding. But at least the native inhabitants of the Canadian Arctic were not to be left without sources of the imported articles that they had formerly obtained from whaling ships and stations, items upon which they had come to depend. Rifles were by this time part of a hunter's standard equipment, but firearms without a supply of ammunition would have been as useless as bows without arrows.

In retrospect, the first two decades of the twentieth century might have been a propitious time for the Canadian government to extend health services and education into arctic communities. Although it did at last endeavour to exert some influence in the Arctic (almost a quarter of a century after receiving the Arctic Islands from Great Britain) the main thrust of its activities was political rather than social. Following the cruise of the steamer *Neptune* in 1903–4 Doctor L.E. Borden had pointed out that there was a clear need for health measures among the Eskimo, but the government did not recognize this as an urgent priority, and preferred to concentrate upon the establishment of police posts. The Eskimos thus became, in the words of the anthropologist Diamond Jenness, "wards of the police" for a time.

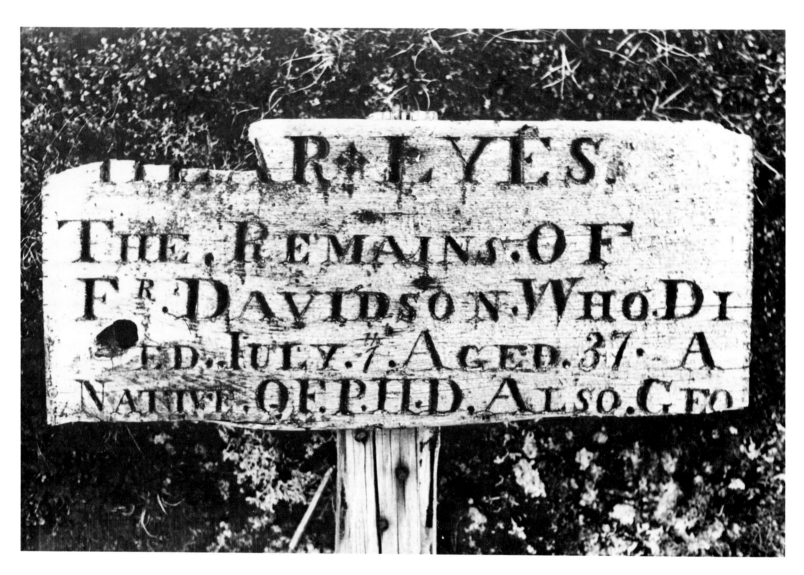

Wooden headboard of a sailor's grave at
Niantelik, near the former Cumberland
Sound whaling station on Blacklead Island.
(Gil Ross)

Epilogue

MORE THAN 3,000 VOYAGES were undertaken to the Davis Strait whaling grounds during the nineteenth and early twentieth centuries. It was an impressive economic operation, involving many thousands of men, the importation of hundreds of thousands of tons of whale oil and baleen into European and American ports, and the diffusion of these raw materials into various branches of industry, in which many other people were kept at work. Whaling was also a magnificent human adventure containing countless moments of courage, endurance, heroism, and achievement, in the face of dreadful conditions of sea, ice, and weather. Important in a socio-political sense as well, whaling expanded the Euro-American knowledge of arctic regions and of their inhabitants, brought the Eskimos of Baffin Island from a subsistance hunting culture based on indigenous materials to a semi-commercial whaling and hunting economy dependent upon imported materials and commodities, and contributed to the extension of British and Canadian sovereignty into the Arctic Islands. The ledger sheet of human achievement might therefore be perceived as having a healthy positive balance.

Yet, from another point of view, the efforts expended in the Davis Strait whale fishery all came to nothing, for by the time of the First World War the whale stock had been destroyed, the industry had faded away, arctic whalemen had become a thing of the past, and the native population of Baffin Island may have declined.

Modern whaling has exploited other species in other oceans. Whale meat has provided human food, stock and poultry feeds, and pet food. Oil, bones, internal organs, ambergris, and stomach contents have been used to make fertilizers, margarine, cooking oils, vitamins, cosmetics, and other products. If things had gone differently there might still be a viable Davis Strait whaling industry, possibly one managed and operated by Canadian Inuit and native Greenlanders on a sustainable-yield basis. But such a possibility does not now exist; the successive extirpation of the bowhead whale by European and American whalemen around Spitsbergen and in the Greenland Sea, Davis Strait, and Hudson Bay left virtually nothing for future generations. The native inhabitants of the Davis Strait region, whose Thule ancestors built a distinctive and admirable culture upon the bowhead

"Right Whale"
Balaena mysticetus —

whale without exterminating it, no longer have the choice of utilizing this species.

Conservationist attitudes and policies, unhappily, are more often born out of scarcity than abundance, with the consequence that they sometimes occur too late. *Balaena mysticetus*, the Greenland or bowhead whale, was eventually granted protection from commercial whaling, but not until 1931, twenty years after the end of whaling. Some whaling captains had spoken up during the nineteenth century, warning that if the killing went on there would soon be no more whales left to kill, but the idea

of a diminished production is often repugnant to industry; and politicians, lacking the long-term view, are seldom inclined to pass measures that threaten to reduce economic activity and employment during their own terms of office, regardless of future benefits. So the arctic whaling industry went blindly on, consuming its own capital at a furious pace until, like the very whales it sought, it perished.

Considering the unfortunate depredations by modern whaling nations upon other species of whales during the last fifty years it may be an impertinence to pass judgement upon the

shipowners, merchants, and politicians of earlier times. But at no point in history can there be any excuse, on a finite earth, for the unthinking destruction of its renewable resources.

As for the whalemen, they never had the capacity to alter the course of events within the industry. Pawns of the economic system, they were merely the employees who did the work of securing oil and bone—and hard work it was. They sailed the ships, manned the whaleboats, plunged the harpoons in, and flayed the dead whales. They routinely endured extraordinary hardship and risk on voyages to the arctic regions in order to earn their livelihood. The narratives that outlast them stand as a tribute to their hardiness and courage.

Left: Bowhead whale with whaleman in foreground. Sketch by R. H. Hilliard. (Glenbow Archives)

Above: Grave headboard at Niantelik. (Gil Ross)

Whale stamp from the journal of Thornton Macklin, surgeon of Narwhal *in 1874. (Dundee Museums)*

Glossary

The nomenclature of masts, yards, rigging, and sails would require a glossary by itself, and for such terms the reader is left to navigate independently. Eskimo words (as used in this book) are given in italics. Words in **bold face** are explained elsewhere in the glossary.

"A fall." The cry given on board a whaler when one of the **whaleboats** was seen to have harpooned a whale. (See **jack**.)

Angakok. Shaman or medicine man.

Angakooeet. The plural of *angakok*.

Angkut. See *angakok*.

Arvik. The bowhead whale *Balaena mysticetus*.

Baleen (also **bone, whalebone**). The 700-odd flexible, hair-fringed slabs (or **fins**) in a bowhead whale's mouth, used to filter out tiny animal food (krill) as sea water is forced out with the tongue.

Bank. A place on deck where skinned blubber pieces were piled before they went to the **choppers**.

Bankers. Men who, during the process of **making-off**, shifted blubber pieces from the **bank** to the chopping blocks.

Bay ice. Newly formed sea ice, which first appears in protected bays.

Beat. To sail to windward. (See **tack**.)

Bearded seal. The large seal *Erignathus barbatus*.

Berg. See **iceberg**.

Bladdernose seal. Hooded seal (the large seal *Cystophora cristata*.)

Blink. See **ice blink**.

Blocks, chopping. Pieces of whales' tails upon which the **choppers** cut up blubber (as we would use a breadboard).

Blubber guy. A rope or hawser extending from foremast to mainmast above the deck, from which **spek** tackles were suspended during making off.

Boatsteerer. In British whalers the man who steered a **whaleboat** during the pursuit and kill of a whale. He ranked second to the **harpooner**, who was in charge of the boat. In American whalers, however, the boatsteerer was also the **harpooner**; after harpooning a whale he went aft to wield the steering oar while the boatheader, or headman of the boat, went forward to take up the lance and make the kill.

Boatswain bird. Probably the parasitic jaeger (*Stercorarius parasiticus* or *S. crepidatus*).

Bollard. The vertical post or "loggerhead" at the bow or stern of a **whaleboat**, around which the whale line was looped to produce friction and create drag behind a harpooned whale. The **whaleboat**, connected to the whale by the line and harpoon, thus acted as a **drogue**. The bollard in British boats was generally at the bow; in American boats it was usually at the stern.

Bone. See **Baleen**.

Brail (also **brail in; brail up**). To gather a loose-footed sail up against the mast, thus spilling out the wind and temporarily furling the sail. This is accomplished by hauling in on small ropes ("brails").

Bran, on the. **Whaleboats** were said to be "on the bran" when cruising leisurely or waiting for whales to appear near the ice.

Brash ice. Small pieces of ice broken away from **floes** or **bergs** by abrasion.

Brig. A two-masted sailing vessel square-rigged on both masts, with a spanker (a fore-and-aft sail) added to the mainmast.

Buntlines. Ropes by which a square sail could be hauled up to the yard, spilling out the wind before furling.

Burgomaster. The glaucous gull (*Larus hyperboreas* or *L. glaucus*).

Butt. A cask holding 126 gallons.

Cant fall. The handling end of the **cant purchase**, hauled in to rotate the whale during **flensing**.

Cant (also **kent**) **piece.** A strip of whale's blubber a few feet wide between its head and fins (where most animals would have a neck), used to hold and rotate the whale carcass during flensing.

Cant (also **kent**) **purchase.** The tackle fixed aloft to the mainmast and extending to the whale's **cant piece.**

Capstan. A ship's device for hauling in on anchor cables, mooring lines, or a whale's blubber during **flensing**. The capstan bars (radiating out from the vertical axle like spokes of a wheel) were walked round by the sailors. Later capstans were powered by steam. Compare **windlass**.

Chain boat. A **whaleboat** carried on davits near the **chains**.

Chains (also **channels**). Horizontal planks extending out from the hull at the level of the upper deck where the mast shrouds descend to their fastenings ("chain plates") on the hull. Also, a position on this plank abreast of the foremast from which sailors swung the lead and line to measure water depth.

Chain plates. See **chains.**

Channels. See **chains.**

Choppers. Men (usually **boatsteerers**) who stood along a long table (the upturned lid of the **spek trough**) cutting blubber into pieces small enough to fall down the **lull** for storage in casks or tanks below decks.

Chopping blocks. See **blocks, chopping.**

Chute. See **lull.**

Clash (also **closh**). A small square platform about three feet high, studded with spikes, mounted on a stanchion. Blubber pieces were lifted onto the spikes with **pick-haaks**, or "clash hooks," for removal of the skin by the **skinners**.

Clashers. Men who placed pieces of blubber on **clashes** for the **skinners**.

Clean (ship). Without a cargo of whale oil; an unsuccessful whaler.

Closh. See **clash.**

Coona. See **cuna.**

Cooney. See **cuna.**

Covering board. The outermost plank of the main deck.

Crang. See **kreng.**

Cuna (also **coona, cooney, cunah, kuna,** etc.). The whaleman's term for an Eskimo woman.

Cunah. See **cuna.**

Deadlight. The hinged metal cover that protects a glass scuttle (circular window) in a ship's side.

Deer. Caribou (*Rangifer tarandus*).

Dovekie. The little auk (*Plautus alle*), a small seabird abundant in arctic waters. Also called **rotje**.

Drag. See **drogue.**

Drogue (also **drag**). A device to create friction or drag in the water behind a ship or a harpooned sea mammal in order to retard its progress.

Duck. Cloth made of linen or cotton, lighter than canvas, and used in outer clothing and light sails. Also, trousers made out of this material.

East Land (also **East Side**). The land bordering the east side of Baffin Bay and Davis Strait; the west coast of Greenland.

"Fall, a." See **"a fall."**

Fall, loose. A pursuit of whales by all a ship's **whaleboats**.

Fall, falls (also **faulds**). The lead rope or handling end of a **tackle**; more generally, the entire tackle, especially those used to lower and hoist **whaleboats** from davits.

Fast ice (also **landfast ice; land floe**). The reasonably stable sea ice that forms along coasts and is fast to the land through the winter.

Faulds. See **falls.**

Fid. A hard, tapered wooden tool used in splicing rope.

Fin (also **fine**). A blade or plate of **baleen**.

Fine. The end part of something, as in a "fine of blubber." Also, occasionally, a **fin** (of **baleen**).

Fin-toe. See **fintow.**

Fintow (also **fin-toe**). A line securing the fins of a dead whale while towing it behind **whaleboats**.

Fish. The whaleman's usual term for a whale. Hence "whale-fishery," "fishing" and so on.

Flense (also **flinch**). To strip off a whale's blubber prior to **making off**.

Flensers. Harpooners who cut the blubber off the whale in the water.

Flinch. See **flense.**

Floe. To the whaleman an extensive field of ice, either attached to the land (land floe) or drifting at sea. Also, (more recently) an individual piece of drifting **pack ice**.

Floe edge. The seaward extremity of the **fast ice**.

Foreganger. A strong rope connecting whale line and **harpoon**.

Frap. A rage (Yorkshire slang).

Gam. The American term for a visit at sea between whaling captains, and sometimes mates, to exchange news. British whalemen called such a meeting a "**mollie**."

Garland. A wreath or similar decoration hoisted to a whaler's masthead on the first of May, during her voyage to the whaling grounds, and removed months later on arrival home, by eager young boys.

Grampus. The killer whale (*Orca gladiator*).

Gum. To clean fragments of a whale's gums from the base of its **baleen** plates.

Guy. See **blubber guy**.

Harpooner. In British whalers the man who had charge of a **whaleboat** and whose job it was to harpoon and lance the whale. In American ships the harpooner (also called **boatsteerer**) was second-in-command of the boat, and after harpooning he had to relinquish the bow position to the boatheader (the boat's chief officer) for the lancing, while he went aft to handle the steering oar. See also **speksioneer**.

Harpooner, loose. A novice **harpooner**.

Hooded seal. See **bladdernose seal**.

Husky. A whaleman's term for Eskimo. See also **Yack**.

Iceberg (also **berg**). A large fragment of land ice broken off, or "calved," from a glacier to drift in ocean waters.

Ice blink. An atmospheric condition in which a low cloud layer reflects the darkness of distant open water, revealing its presence among or beyond sea ice.

Igloo (also *iglu*). An Eskimo dwelling, most commonly the snow house of winter.

Iglu. See *igloo*.

Innuit. See *Inuit*.

Inuit (*Innuit*). The Eskimo word for themselves; "the people."

Jack. The flag signal displayed by a **whaleboat** as soon as it was fast to a whale. Also the flag hoisted to the head of

a whaler's mizzenmast to signify that one of her boats was fast.

Kabluna (also *kadluna*, *Kudlunite*, etc.). Eskimo word for a white person.

Kadluna. See *Kabluna*.

Kayak (also *kyak*). The narrow skin boat used by a hunter in coastal waters and on lakes. It was decked in, with an open cockpit which usually accommodated a single person.

Kent piece. See **cant** piece.

Kent purchase. See **cant** purchase.

Kings. Men in the hold who packed blubber into casks during **flensing**, and later heaved the blubber up onto deck for **making-off**.

Komatik. The long travelling sled pulled by dogs.

Koutak. Stone hammer.

Kreng (also **crang**). A whale carcass after **flensing**.

Krengers. Men on deck who received the blubber sent up by the **kings** for **making-off**, and then removed the stringy parts before it went to the **skinners**.

Kudlinite. See *Kabluna*.

Kummatik. See *komatik*.

Kuna. See **cuna**.

Kyak. See *kayak*.

Landfast ice. See **fast ice**.

Land floe. See **fast ice**.

Larboard (also **port**). A vessel's left-hand side when looking forward. The opposite of **starboard**.

Log. A revolving propeller that recorded the distance travelled by a ship through the water during a day. Also a short version of **logbook**.

Logbook. The official day-to-day record of a ship's voyage. Also called a "log."

Loggerhead. See **bollard**.

Log, sea. A ship's **logbook** on an ocean

passage, during which days were reckoned from noon to noon instead of midnight to midnight. The nautical day of the sea log commenced twelve hours earlier than the normal calendar day.

Loom. The common or Atlantic murre – or guillemot (*Uria aalge*) – or more likely the thick-billed murre – or Brunich's guillemot (*Uria lomvia*).

Loomery. A cliff nesting place of **looms**.

Loose fall. See **fall, loose**.

Loose harpooner. See **harpooner, loose**.

Lull. A canvas hose about a foot in diameter through which chopped-up blubber was conveyed to casks below deck during **making-off**.

Lull nippers. Wooden sticks controlling the movement of blubber pieces down the **lull**. When a cask was full the "lull boy" was ordered to "nip the lull."

Make off. To cut up a whale's **flensed** blubber and store it in casks below.

Mallemuck. The fulmar petrel (*Fulmar glacialis*). Popularly called a **mollie**.

Maulie. See **mollie**.

Meridian. A line of longitude. The time at which the sun crosses an observer's meridian is considered local noon; hence meridian may mean "noon."

Middle Ice. The **pack ice** between the **East Side** and the **West Side**, in the central parts of Baffin Bay and Davis Strait.

Milldolling. Breaking young ice ahead of a ship by rolling a **whaleboat**, suspended from the bowsprit, from side to side, or by dropping a heavy object repeatedly from the bowsprit.

Miss stays. See **stays, miss**.

Molley. See **mollie**.

Mollie (also **molley**, **maulie**, etc). Popular version of **mallemuck**.

Mollie boats. **Whaleboats** tied alongside the whale during **flensing** to accommodate the knives and spades of the flensers working on the whale. (Named for the flocks of **"mollies"** or fulmars attracted to the carcass).

Nessak. Hood of garment.

Nipped. Squeezed by **pack ice** movement.

Nips, in the. A vessel undergoing a "nip," or squeeze, from ice.

Nixir. Hook.

Northeast Water. Possibly a synonym for the **North Water**, or its easternmost extension.

North Water. The perennial open-water area (**polynia**) that persists in winter north of the **Middle Ice** of Baffin Bay but south of Smith Sound.

Nose tackle. A strong rope holding the head of a whale towards the stern of a whaler during **flensing**. Compare **rump rope**.

On the bran. See **bran, on the**.

Ooloo. See *ulu*.

Oomiak. See *umiak*.

Pack ice. Innumerable sea ice fragments of various sizes, drifting with winds and surface currents. Also called "drift ice."

Panniers. Framework holding out the **port** and **starboard** sides of a woman's skirt, as a bustle held out its stern.

Parbuckle. To hoist an object by means of a looped rope, usually with its bight around a stanchion or other fixture.

Pickaniny (also **pikaniny, pikininy**, etc.). The whaleman's term for an Eskimo child.

Pick-haak (also **picky-haak, pick-hook, picky**, etc.). A tool resembling a boathook, with sharper points, used to place blubber pieces upon the **clash** for removal of the skin.

Pick-hook. See **pick-haak**.

Picky. See **pick-haak**.

Picky-haak. See **pick-haak**.

Pikaniny. See **pickaniny**.

Pikininy. See **pickaniny**.

Pilk (also **pilch**). To scrape or pick.

Pilch. See **pilk**.

Ply. To sail back and forth in leisurely fashion without a specific course or destination.

Polynia. An area remaining ice-free in polar waters, even in winter.

Port. See **larboard**.

Reach. To sail obliquely downwind (with the wind on the quarter).

Reef. To reduce sail area in strong winds by tying the lower part of a sail down to the boom. This is accomplished by securing "reef points" (short cords fixed along the sail at intervals) with "reef knots."

Ringtail. An extension to the after end of a fore-and-aft sail, which increases its area.

Rocket. An explosive projectile fired at a whale from the harpoon gun mounted on a **whaleboat**; unlike a harpoon it had no line attached, the object being to kill the whale rather than to connect it to the boat.

Rocknosing. Cruising close to the land in **whaleboats**, mainly to intercept whales migrating southward along the coast of Baffin Island in late summer.

Roach, roachie. See **rotje**.

Rodge. See **rotje**.

Rotch, rotchie. See **rotje**.

Rotje (also **roach, roachie, rotch, rotchie, rodge**, etc.). The dovekie or little auk (*Plautus alle* or *Alea alle*).

Rounders. A game played with bat, ball, and bases, similar to baseball.

Rowraddy. A canvas harness worn by whalemen when tracking a vessel along the **floe edge**.

Rugging. Pulling, tugging, or seizing hold of.

Rump rope. A strong rope holding the tail of a whale towards the bow of a vessel during **flensing**. Compare **nose tackle**.

Sailing ice. Scattered **pack ice** through which a vessel could sail.

Sally. To rock a vessel out of sea ice by having the crew run in unison from side to side.

Scaffer. One who eats voraciously.

Sconce. A small field of pack ice (smaller than a **floe**). Also its individual components, or "sconce pieces."

Scuttlelass. A "scuttlebutt," or water cask for daily use on board ship.

Sea horse. The walrus (*Odobenus rosmarus*).

Seal. Usually either ringed seal (*Phoca hispida*) or harp seal (*Phoca groenlandica*). The whalemen seldom distinguished between these two species, although they did note the difference between **bearded seal** and **bladdernose seal**.

Sea log. See **log, sea**.

Shakes. Barrel staves carried on a whaler and made up into barrels as needed.

Shore lead. A strip of open water between the **fast ice** along the coast and the **pack ice** beyond the **floe edge**.

Sinnet. Mats or pads of interwoven (plaited) rope yarn used mainly to prevent the rigging from chafing the sails.

Sizefish. One in which the longest **baleen** slab was at least six feet. An undersize fish had **baleen** shorter than six feet. (Despite this distinction whalers often killed small whales, for their oil.)

Skeeman. The officer in charge of operations below deck during the process of **making off**.

Skinners. Men (usually **harpooners**) who, during the process of **making-off**, cut the skin from blubber pieces stuck on **clashes**.

Snowbird. Ivory gull (*Pagophila eburnea*).

Southwest Fishery. The whaling carried out in April, May, and June at the margin of the **Southwest Ice**.

Southwest Ice. The **pack ice** off southeastern Baffin Island, the mouth of Hudson Strait, and northern Labrador.

Span (also **span on, span in**). To splice a harpoon to its **foreganger**.

Speak. To communicate with a ship.

Spek. Blubber.

Speksioneer. The officer in charge of the **flensing** operation, normally the chief **harpooner** (on British whalers).

Spek tackles. **Tackles** used to hoist strips of blubber from whales being **flensed**.

Spek trough. A long wooden trough which received small blubber pieces from the chopping blocks before they were sent below deck through the **lull**.

Split. A slab or plate of **baleen**. Sometimes called a **fin**.

Spurs. Spikes worn beneath the boots of the flensers while working on a whale carcass to keep from slipping. (Similar to the crampons of a mountaineer.)

Starboard. The right-hand side of a vessel when looking forward. The opposite of port, or **larboard**.

Stays, in. A sailing vessel is said to be **in stays** when she is pointed up into the wind and her sails are not filled.

Stays, miss. A sailing vessel is said to **miss stays** when, in the act of coming about from one tack to the other, she gets **in stays** and falls back onto the original tack, failing to complete the manoeuvre.

Stove. Broken in, burst, holed (usually referring to a boat).

Tack. To sail into the wind on a course, or **tack**, which is oblique to the wind direction. Or to beat to windward in zig-zag fashion with a series of tacks. Also, to come about from one tack to another by turning the bow of the vessel across the direction of the wind (the opposite of **wear**).

Tackle. A purchase in which blocks (pulleys) and ropes provide mechanical advantage. Used on board whalers in **flensing**, lowering and hoisting boats, and many other tasks.

Taffarel. See **taffrail**.

Taffrail (also **taffarel**). A rail running around the deck at the stern of a ship.

Tarngnil. Soul.

Toopick. See *tupik*.

Toopik. See *tupik*.

Tongak. Spirit.

Tougak. See *tongak*.

Troak. See **truck**.

Truck (also **troak**). To barter (with the Greenland and Canadian Eskimos). Also the articles obtained in trade.

Tupik (also *toopik*, *toopick*). Tent.

Tupok. See *tupik*.

Tupuyck. See *tupik*.

Ulu (also *ooloo*). The traditional woman's knife with crescent-shaped blade.

Umiak (also *oomiack*). The large, open, skin boat used for travelling, freighting, and sometimes whaling.

Unicorn (also **unie**). The narwhal (*Monodon monoceros*).

Ursuk. The **bearded seal**.

Waist boat. A whaleboat carried in the midship section of a whaler's upper deck between the foremast and mainmast.

Warp. To change the position of a ship by adjusting "warps" (ropes or hawsers) attached to the shore, the ice, or a kedge anchor.

Watery poops. Possibly a sort of individual slang for "water proofs."

Wear. To jibe or bring the wind from one side of a sailing vessel to the other by turning the stern across the wind, rather than the bow across as in **tacking**.

West Ice (also **West Pack**). A term sometimes used along the Greenland coast to refer to the **Middle Pack** or more specifically to pack ice lying along the coast of Baffin Island.

West Land (also **West Side**). The land bordering the west side of Baffin Bay and Davis Strait; the east coast of Baffin Island.

Whaleboat. An open boat twenty-five to thirty feet long from which whales were harpooned, lanced, and towed to the whaleship, usually by a crew of six men.

Whalebone. See **baleen**.

White stocking day. A day during the month on which whalemen's wives or dependents would go to the shipping office to collect their allowance while the men were away whaling.

Winding sheet. The burial wrapping of a corpse.

Windlass. Like the **capstan**, a contrivance for hauling in on ropes or cables under strain, but the axle is horizontal rather than vertical.

Yack. (also **Yak**, **Yakkie**, etc.). The whaleman's term for an Eskimo. See also **Husky**.

Yak. See **Yack**.

Yakkie. See **Yack**.

Note on Sources

THE NARRATIVES

CHAPTER 1.

Cass, William Eden. "Hull whaler *Brunswick*. A journal of a voyage to Davis's Straits and Baffin's Bay with a brief description of those regions and the whale fishery," 1824. Typescript of manuscript journal. Hull: Museums and Art Galleries. (Another version of this journal—MG24H69 — exists in the Public Archives of Canada, Ottawa, and an abridged version of this is in Goole Library, Humberside, England).

CHAPTER 2.

Duncan, David. 1827. *Arctic regions: voyage to Davis' Strait, by David Duncan, master of the ship Dundee....* London: E. Billing.

CHAPTER 3.

Anonymous. 1831. Manuscript journal. Aberdeen: Department of Manuscripts and Archives, Aberdeen University Library, King's College (MS 673).

CHAPTER 4.

Wanless, John. 1834. "Journal of a voyage to Baffins Bay ship *Thomas* commanded by Alex Cook." Manuscript journal. Dundee: Museum and Art Galleries Department (MS Bo. GB/57/423).

CHAPTER 5.

Elder, William. 1835–36. "Memorandum book." Manuscript journal. Cambridge: Scott Polar Research Institute (MS 823.1, 2).

CHAPTER 6.

Gibb, David. 1837. *A narrative of the sufferings of the crew of the Dee while beset in the ice at Davis' Straits, during the winter of 1836...* Aberdeen: Geo. Clark & Son.

CHAPTER 7.

M'Donald, Alexander. 1841. *A narrative of some passages in the history of Eenoolooapik, a young Esquimaux who was brought to Britain in the ship 'Neptune' of Aberdeen.* Edinburgh: Fraser & Hogg.

CHAPTER 8.

Goodsir, Robert Anstruther. 1850. *An arctic voyage to Baffin's Bay and Lancaster Sound, in search of friends with Sir John Franklin.* London: John Van Voorst.

CHAPTER 9.

Lithgow, Stewart A. 1853. Manuscript journal. Edinburgh: National Library of Scotland.

CHAPTER 10.

Whitehouse, Albert Johnston. 1859–60. Manuscript journal. Hull: Museums and Art Galleries.

CHAPTER 11.

Sullivan, John. 1861. Written statement drawn up before a magistrate and later published in: Hall, Charles Francis, 1865. *Life with the Esquimaux....* London: Sampson, Low, Son, & Marston, p. 73–77.

CHAPTER 12.

Blake, E. Vale. (ed.) 1874. *Arctic experiences: containing Capt. George E. Tyson's wonderful drift on the ice floe....* New York: Harper & Brothers.

CHAPTER 13.

Macklin, Thomas Thornton. 1874. "Notes on a whaling voyage to Davis Straits in the s.s. *Narwhal* of Dundee." Manuscript journal. Dundee: Museums and Art Galleries Department.

CHAPTER 14.

Livingstone-Learmonth, Walter. 1889. Manuscript journal. Ottawa: Public Archives of Canada (MG 29, B 28, vol. 1).

CHAPTER 15.

Cardno, David. n.d. Manuscript diaries. Aberdeen: Department of Manuscripts and Archives, Aberdeen University Library, King's College (MS 3090/7).

THE ILLUSTRATIONS

Grateful acknowledgement is made to the following archives and other illustration sources for granting permission to use material in their possession: Aberdeen University Library, Aberdeen, Scotland; Center for Polar and Scientific Archives, National Archives and Records Administration, Washington, D.C.; Dundee Museums and Art Galleries, Dundee, Scotland; Glasgow University Library, Glasgow, Scotland; Glenbow-Alberta Institute, Calgary, Alberta; Houghton Library, Harvard University, Cambridge, Massachusetts; Humberside County Council, Central Library, Hull, England; Town Docks Museum; City of Kingston Upon Hull Museums and Art Galleries, Hull, England; Kendall Whaling Museum, Sharon, Massachusetts; Provincial Archives, Winnipeg, Manitoba; Metropolitan Toronto Library Board, Toronto, Ontario; The MIT Museum, Cambridge, Massachusetts; National Film Board of Canada, Phototèque, Ottawa, Ontario; Trustees of the National Library of Scotland, Edinburgh, Scotland; National Maritime Museum, London, England; Peabody Museum, Salem, Massachusetts; Peary-MacMillan Arctic Museum and the Arctic Studies Center, Bowdoin College, Brunswick, Maine; Public Archives of Canada, National Photography Collection, Ottawa, Ontario; Scott Polar Research Institute, Cambridge, England; Mr. and Mrs. William Cruden, Aberdeen, Scotland.

Suggestions for Further Reading

ADLARD, MARK. *The Greenlanders* (Harmondsworth, Middlesex: Penguin 1980). A colourful and gripping novel of the British whale fishery east of Greenland, full of details of whaling and life on board ship.

BOCKSTOCE, JOHN R. *Steam whaling in the Western Arctic*. With contributions by William A. Baker and Charles F. Batchelder (New Bedford, Massachusetts: Old Dartmouth Historical Society 1977). Well written, accurate, and based on original research, this profusely illustrated book describes late nineteenth century whaling around Alaska, eastern Siberia, and Canadian waters of the Beaufort Sea.

DYKES, JACK. *Harpoon to kill* (London: J.M. Dent 1972). A colourful novel based on the ice-drift of the *Diana* in 1866–67 (see Smith below).

FERGUSON, ROBERT. *Arctic harpooner; a voyage on the schooner Abbie Bradford* 1878–1879. Edited by Leslie Dalrymple Stair (Philadelphia: University of Pennsylvania Press 1938). A heavily edited but highly readable account by an American whaleman wintering in Hudson Bay, the second of the three major whaling grounds exploited in northern Canada.

FRANCIS, DANIEL. *Arctic chase: a history of whaling in Canada's north* (St. John's, Newfoundland: Breakwater Books 1984). An overview of whaling on the Davis Strait, Hudson Bay, and Beaufort Sea grounds.

JENKINS, JAMES TRAVIS. *A history of the whale fisheries* (Port Washington, N.Y.: Kennikat Press 1971). A re-issue of a long-respected work, first published in 1921, and valuable for its historical and geographical sweep. It covers European and American whaling from the tenth century to the early twentieth.

JACKSON, GORDON. *The British whaling trade* (Hamden, Connecticut: Archon Books 1978). A valuable economic history of British whaling from the seventeenth century to the mid-twentieth century, which includes several chapters on the Arctic.

LINDSAY, DAVID MOORE. *A voyage to the Arctic on the whaler Aurora* (Boston: Dana Estes 1911). A firsthand narrative of a Davis Strait voyage on a Dundee steam whaler in 1884 written by the ship's surgeon, a medical student at the University of Edinburgh.

LUBBOCK, BASIL. *The arctic whalers* (Glasgow: Brown, Son, & Ferguson 1955). Based largely on the work of John Suddaby of Hull, and conveniently organized on a year-to-year basis, this book is enlivened by Lubbock's acquaintance with seafaring men and his wide knowledge of sailing ships, and is generously illustrated by paintings and photographs (including a number by Livingstone-Learmonth). The scholar should beware of factual errors and inconsistencies, however.

MACKINTOSH, A.W. *A whaling cruise in the arctic regions* (London: Hamilton, Adams 1889). An autobiographical account of a Davis Strait voyage, by yet another medical man, on the Dundee steam whaler *Active* in 1877.

MARKHAM, ALBERT HASTINGS. *A whaling cruise to Baffin's Bay and the Gulf of Boothia....* (London: Sampson, Low, Marston, Low, and Searle

1875). The experiences of a renowned authority and writer on arctic exploration, who in 1873 accompanied the Dundee steam whaler *Arctic*, the vessel that surgeon Macklin saw destroyed in Prince Regent Inlet in the following year (chapter 13).

ROSS, W. GILLIES. *Whaling and Eskimos: Hudson Bay 1860–1915* (Ottawa: National Museum of Man, Publications in Ethnology, no. 10 1975). A geographical and ethnographical analysis of the American-dominated Hudson Bay fishery and its impact on the native population.

ROSS, W. GILLIES (ed.). *An arctic whaling diary; the journal of Captain George Comer in Hudson Bay 1903–1905* (Toronto: University of Toronto Press 1984). A detailed journal by the best-known of Hudson Bay whaling masters, who in addition to whaling responsibilities recorded ethnographical information about the Eskimos of various tribes, made sound recordings of their songs, and collected samples of their tools, weapons, and clothing for museums.

SCORESBY, WILLIAM, JR. *An account of the arctic regions with a history and description of the northern whale-fishery.* 2 vols. (Newton Abbot, Devon: David & Charles 1969). A facsimile edition of the classic work on whaling and the arctic regions, originally published in 1820 and still a standard reference, as well as fascinating reading.

SMITH, CHARLES EDWARD. *From the deep of the sea: the diary of Charles Edward Smith surgeon of the whale-ship* DIANA, *of Hull.* Edited by Charles Edward Smith Harris (Edinburgh: Paul Harris 1977). A reprint of one of the most widely read books about arctic whaling, describing the ice-drift of the ship in Baffin Bay and Davis Strait, in 1866–67. The valiant efforts of surgeon Smith to save the men and the ship following the death of the captain in December were later rewarded by testimonials and gifts presented by the grateful people of Hull.

Index

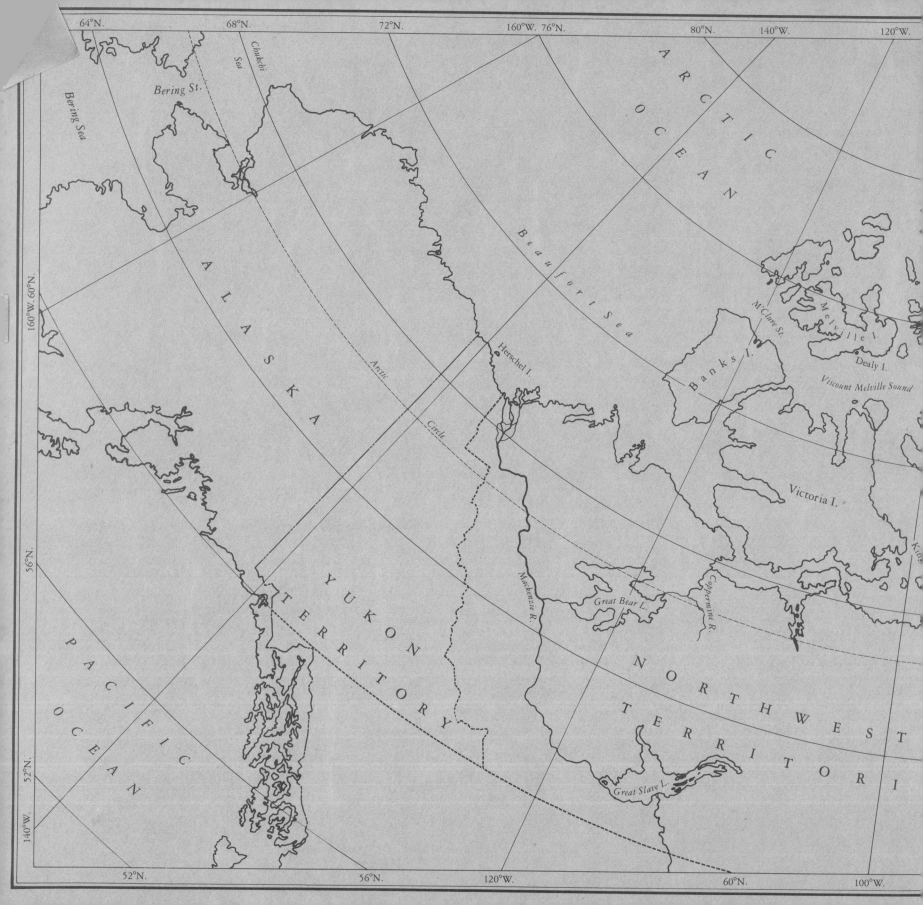